SpringerBriefs in Mathematics

SpringerBriefs in Mathematics showcases expositions in all areas of mathematics and applied mathematics. Manuscripts presenting new results or a single new result in a classical field, new field, or an emerging topic, applications, or bridges between new results and already published works, are encouraged. The series is intended for mathematicians and applied mathematicians.

For further volumes:
http://www.springer.com/series/10030

BCAM SpringerBriefs

Editorial Board

BCAM *SpringerBriefs* aims to publish contributions in the following disciplines: Applied Mathematics, Finance, Statistics and Computer Science. BCAM has appointed an Editorial Board that will evaluate and review proposals.

Typical topics include: a timely report of state-of-the-art analytical techniques, bridge between new research results published in journal articles and a contextual literature review, a snapshot of a hot or emerging topic, a presentation of core concepts that students must understand in order to make independent contributions.

Please submit your proposal to the Editorial Board or to Francesca Bonadei, Executive Editor Mathematics, Statistics, and Engineering: francesca.bonadei@ springer.com

Sylvain Ervedoza • Enrique Zuazua

Numerical Approximation of Exact Controls for Waves

basque center for applied **mathematics**

Sylvain Ervedoza
Institut de Mathématiques
  de Toulouse & CNRS
Université Paul Sabatier
Toulouse, France

Enrique Zuazua
BCAM-Basque Center for Applied
  Mathematics
Ikerbasque, Bilbao
Basque Country, Spain

ISSN 2191-8198   ISSN 2191-8201 (electronic)
ISBN 978-1-4614-5807-4   ISBN 978-1-4614-5808-1 (eBook)
DOI 10.1007/978-1-4614-5808-1
Springer New York Heidelberg Dordrecht London

Library of Congress Control Number: 2012956551

Mathematics Subject Classification (2010): 35L05, 93B05, 93B07, 34H05, 35L90, 49M25

Printed on acid-free paper

Springer is part of Springer Science+Business Media (www.springer.com)

Preface

In this book, we fully develop and compare two approaches for the numerical approximation of exact controls for wave propagation phenomena: the continuous one, based on a thorough analysis of the continuous model, and the discrete one, which relies upon the analysis of the discrete models under consideration. We do it in the abstract functional setting of conservative semigroups.

The main results of this book end up unifying, to a large extent, these two approaches yielding similar algorithms and convergence rates. The discrete approach, however, has the added advantage of yielding not only efficient numerical approximations of the continuous controls but also ensuring the partial controllability of the finite-dimensional approximated dynamics, i.e., the fact that a substantial projection of the approximate dynamics is controlled. It also leads to iterative approximation processes that converge without a limiting threshold in the number of iterations. Such a threshold has to be taken into account, necessarily, for methods derived by the continuous approach, and it is hard to compute and estimate in practice. This is a drawback of the methods emanating from the continuous approach that exhibit divergence phenomena when the number of iterations in the algorithms aimed to yield accurate approximations of the control goes beyond this threshold.

We shall also briefly explain how these ideas can be applied for data assimilation problems.

Though our results apply in a wide functional setting, our approach requires a fine analysis in the case of unbounded control operators, e.g., in the case of boundary controls. We will therefore show how this can be done in a simple case, namely the $1 - d$ wave equation approximated by finite-difference methods. In particular, we present several new results on the rates of convergence for the solution of the wave equation with nonhomogeneous Dirichlet boundary data.

Toulouse, France   Sylvain Ervedoza
Bilbao, Basque Country, Spain   Enrique Zuazua

Acknowledgments

Sylvain Ervedoza is partially supported by the Agence Nationale de la Recherche (ANR, France), Project C-QUID number BLAN-3-139579, Project CISIFS number NT09-437023, the AO PICAN of University Paul Sabatier (Toulouse 3), and grant MTM2011-29306 of the MICINN, Spain. Part of this work has been done while he was visiting the BCAM—Basque Center for Applied Mathematics, as a visiting fellow.

Enrique Zuazua is supported by the ERC Advanced Grant FP7-246775 NU-MERIWAVES, the Grant PI2010-04 of the Basque Government, the ESF Research Networking Program OPTPDE, and Grant MTM2011-29306 of the MICINN, Spain.

Contents

Introduction

Motivation

Let Ω be a smooth bounded domain of $\mathbb{R}^n$ and consider an open subset $\omega \subset \Omega$.
We consider the controlled wave equation in Ω:

$$\begin{cases} \partial_{tt} y - \Delta y = v\chi_\omega, & (t,x) \in \mathbb{R}_+ \times \Omega, \\ y = 0, & (t,x) \in \mathbb{R}_+ \times \partial\Omega, \\ (y(0,x), \partial_t y(0,x)) = (y_0(x), y_1(x)), \, x \in \Omega. \end{cases} \quad (1)$$

Here, y, the state of the system, may represent various wave propagation phenomena as, for instance, the displacement for elastic strings and membranes or acoustic waves. The control function is represented by v which is localized in the control subdomain ω through χ_ω, the characteristic function of ω in Ω.

This work is devoted to discuss, analyze, and compare two approaches for the numerical approximation of exact controls, the continuous and discrete ones.

System (1) is said to be exactly controllable in time T if, for all $(y_0, y_1) \in H_0^1(\Omega) \times L^2(\Omega)$ and $(y_0^T, y_1^T) \in H_0^1(\Omega) \times L^2(\Omega)$, there exists a control function $v \in L^2((0,T) \times \omega)$ such that the solution y of Eq. (1) satisfies

$$(y(T), \partial_t y(T)) = (y_0^T, y_1^T). \quad (2)$$

Such property is by now well known to hold under suitable geometric conditions on the set ω in which the control is active, the domain Ω in which the equation is posed, and the time T during which the control acts.

In the seminal work of Lions [36], in which the Hilbert uniqueness method (HUM) was introduced, the problem was reduced to that of the observability of the adjoint system and multiplier methods were derived for the latter to be proved under suitable geometric restrictions (see also [27, 31] for other types of multipliers). Later, in [3, 5] it was shown that system (1) is exactly controllable in time $T > 0$ if and only if (ω, Ω, T) satisfies the so-called geometric control condition (GCC). Roughly speaking, this condition states that all rays of geometric optics—

which in the present case are straight lines bouncing on the boundary $\partial\Omega$ according to Descartes law—should enter into the control subset ω in a time less than T.

All along this work we shall assume that (ω, Ω, T) fulfills the GCC. In that case, for all (y_0, y_1), (y_0^T, y_1^T) in $H_0^1(\Omega) \times L^2(\Omega)$, the existence of a control function $v \in L^2((0, T) \times \omega)$ such that the corresponding solution of Eq. (1) satisfies Eq. (2) is guaranteed.

The question we address is that of building efficient numerical algorithms to compute such a control.

Control and Numerics

Of course, this problem is not new, and many articles have been devoted to it.

In the pioneering works [21–23] (see also the more recent book [24]) it was shown that high-frequency spurious solutions generated by the discretization process could make the discrete controls diverge when the mesh size goes to zero. These results have later received a thorough theoretical study (see, for instance, [28] in which the finite-difference and finite element methods in $1 - d$ on uniform meshes were addressed and the more recent survey articles [16, 52]).

The analysis developed in these articles leads to the necessity of distinguishing two different approaches, the continuous and the discrete ones. In the continuous one, after characterizing the exact controls of the continuous wave equations, the emphasis is placed on building efficient numerical methods to approximate them. In the discrete one, by the contrary, one analyzes the controllability of discrete models obtained after discretizing the wave equation by suitable numerical methods and their possible convergence towards the controls of the continuous models under consideration when the mesh-size parameters tend to zero.

In other words, to compute approximations of controls for continuous models, there are mainly two alternative paths:

first CONTROL and then NUMERICS

or

first NUMERICS and then CONTROL.

In this book we first focus on the continuous approach, the key point being to build an iterative process in an infinite-dimensional setting yielding the control of the continuous wave equation, to later approximate it numerically. To be more precise, we approximate numerically each step of this iterative process. Of course, this generates error terms in each iteration that add together and eventually may produce divergence phenomena, when the number of iterations goes beyond a threshold.

One of the most natural manners to derive such an iterative algorithm is in fact the implementation of the HUM method that characterizes the control of minimal norm, by minimizing a suitable quadratic functional defined for the solutions of the adjoint system. The minimizer can then be approximated by gradient descent algorithms. This leads naturally to an iterative algorithm to compute the control of the continuous model that later can be approximated by standard numerical approximation methods, such as finite differences and elements.

Recently, a variant of this continuous approach has been developed in [9] following Russell's technique [47] to construct the control out of stabilization results. According to Russell's approach, the control can be built as the fixed point of a contractive map, whose contractivity is ensured by the stabilizability of the system. This leads then naturally to an iterative method for approximating a continuous control. Note, however, that the control obtained in this manner is not the one of minimal norm (the one given by HUM) but rather that obtained through Russell's *stabilization implies control principle*. A similar method has been numerically implemented successfully in [1] in the context of data assimilation problems for some nonlinear models as well.

As we shall see, once the iterative algorithm that the continuous approach yields is projected into the finite-dimensional numerical approximation models, we end up with a method that is very similar, in form and computational cost, to the one obtained by means of the discrete approach. The latter consists of building discrete approximation models whose controls converge to the one of the continuous dynamics usually after filtering the spurious numerical components.

The first main advantage of the discrete approach is that it yields approximate controls that control, at least partially, the approximated numerical dynamics. But this is done to the price of carefully analyzing the control properties of the finite-dimensional dynamics, an extra and often complicated task that is not required when developing continuous methods. As we shall explain, developing the discrete approach is also computationally relevant since it allows to use much faster iterative algorithms. The continuous approach is conceptually simpler, however. Indeed, it superposes the continuous control theory to build an iterative algorithm in the continuous setting and classical numerical analysis to approximate it effectively, without getting involved into fine controllability properties of the discrete dynamics.

The results we shall present below apply in the much more general setting of conservative semigroups, for which the wave equation (1) is the most paradigmatic example. Most of the presentation will then be done in this abstract unifying frame.

Our main results on the comparison of both approaches in the abstract setting are presented in Chap. 1.

On the Convergence of the Numerical Schemes

Though the results of Chap. 1 apply in a very general setting, one of our main applications is the boundary control $1 - d$ wave equation discretized using finite-difference (or finite element) methods; see Sect. 1.7. In such case, the unboundedness of the control operator makes it hard to check the convergence assumptions of Chap. 1.

We therefore provide a fine analysis of the convergence properties of finite-difference methods that do not seem to be available in the existing literature. Thus in Chaps. 3 and 4 we develop some new technical results on the convergence of the finite-difference approximation methods for the wave equation and, in particular, on nonhomogeneous boundary value problems that are necessary for a complete analysis of the convergence of numerical controls towards continuous ones. These results are of interest independently of their control theoretical implications.

The main difficulty to obtain convergence rates for numerical approximations is that solutions of the (even in $1-d$) wave equation with nonhomogeneous boundary data are defined in the sense of *transposition*.

To be more precise, following [36] (see also [33, 35]), if v belongs to $L^2(0,T)$, the solution y of

$$\begin{cases} \partial_{tt}y - \partial_{xx}y = 0, & (t,x) \in \mathbb{R}_+ \times (0,1), \\ y(t,0) = 0, y(t,1) = v(t), & t \in \mathbb{R}_+, \\ (y(0,x), \partial_t y(0,x)) = (0,0), x \in (0,1), \end{cases} \tag{3}$$

in the sense of transposition lies in $C([0,T];L^2(0,1)) \cap C^1([0,T];H^{-1}(0,1))$.

The proof of this fact is based on a hidden regularity (or admissibility) result for the solutions φ of the adjoint system

$$\begin{cases} \partial_{tt}\varphi - \partial_{xx}\varphi = f, & (t,x) \in (0,T) \times (0,1), \\ \varphi(t,0) = \varphi(t,1) = 0, & t \in (0,T), \\ (\varphi(T,x), \partial_t\varphi(T,x)) = (0,0), x \in (0,1), \end{cases} \tag{4}$$

with source term $f \in L^1(0,T;L^2(0,1))$ (and for $f = \partial_t g$ with $g \in L^1(0,T;H_0^1(0,1)))$, which should satisfy

$$\partial_x \varphi(t,1) \in L^2(0,T). \tag{5}$$

Note that, with these regularity assumptions on the initial data and the source term, solutions φ of Eq. (4) belong to the space $C([0,T];H_0^1(0,1)) \cap C^1([0,T];L^2(0,1))$, but this well-known finite energy property does not guarantee Eq. (5) to hold by classical trace inequalities. In fact, Eq. (5) is a consequence of a fine property of hidden regularity of solutions of the wave equation with Dirichlet boundary conditions, both in the $1-d$ and in the multidimensional case. Thus, for the analysis of the convergence of the numerical approximation methods these hidden regularity' properties have to be proved uniformly with respect to the mesh-size parameters.

Hence the sharp analysis of the convergence of the finite-difference approximations of the solutions of Eq. (3) will be achieved in two main steps:

- In Chap. 2 we study the behavior of the finite-difference approximation schemes of Eq. (4) from the point of view of admissibility. In particular, we prove a uniform admissibility result (already obtained in [28]) that will be needed for the convergence results. Our proof relies on a discrete multiplier technique. We also explain how this can be used to obtain sharp quantitative estimates for a uniform observability result within classes of filtered data.

- In Chap. 3 we present the convergence of the $1-d$ finite-difference approximation schemes with homogeneous Dirichlet boundary data and establish sharp results about convergence rates. Most of these results are rather classical, except for the convergence of the normal derivatives.
- In Chap. 4 we derive convergence results for the finite-difference approximation on the $1-d$ wave equation (3) with nonhomogeneous boundary data, based on suitable duality arguments.

Further Comments

In Chap. 5, we conclude our study with some further comments and open problems. In particular, we comment on the consequences of our analysis at the level of optimal control problems or the extension of our results to the fully discrete context.

Chapter 1
Numerical Approximation of Exact Controls for Waves

1.1 Introduction

We present an abstract framework in which our methods and approach apply, the wave equation being a particular instance that we present in Sect. 1.7.

1.1.1 An Abstract Functional Setting

Let X be an Hilbert space endowed with the norm $\|\cdot\|_X$ and let $\mathbb{T} = (\mathbb{T}_t)_{t \in \mathbb{R}}$ be a linear strongly continuous group on X, with skew-adjoint generator $A : \mathscr{D}(A) \subset X \to X$, satisfying $A^* = -A$. We shall also assume that A has compact resolvent and that the domain of A is dense in X.

For convenience, we also assume that 0 is not in the spectrum of A, so that for $s \in \mathbb{N}$, we can define the Hilbert spaces $X_s = \mathscr{D}(A^s)$ of elements of X such that $\|A^s x\|_X < \infty$ endowed with the norm $\|\cdot\|_s := \|A^s \cdot\|_X$. Note that this does not restrict the generality of our analysis. Indeed, if 0 is in the spectrum of A, choosing a point $\beta \in i\mathbb{R}$ which is not in the spectrum of A and replacing A by $A - \beta I$, our analysis applies.

For $s \geq 0$, we also define the Hilbert spaces X_s obtained by interpolation between $\mathscr{D}(A^{\lfloor s \rfloor})$ and $\mathscr{D}(A^{\lceil s \rceil})$, which we endow with the norm $\|\cdot\|_s$. For $s \leq 0$, we then define X_s as the dual of X_{-s} with respect to the pivot space X and we endow it with its natural dual norm.

We are then interested in the following equation:

$$y' = Ay + Bv, \quad t \geq 0, \qquad y(0) = y_0 \in X. \tag{1.1}$$

Here, B is an operator in $\mathscr{L}(U, X_{-1})$, where U an Hilbert space. This operator determines the action of the control function $v \in L^2_{\text{loc}}([0, \infty); U)$ into the system.

S. Ervedoza and E. Zuazua, *Numerical Approximation of Exact Controls for Waves*,
SpringerBriefs in Mathematics, DOI 10.1007/978-1-4614-5808-1_1,
© Sylvain Ervedoza and Enrique Zuazua 2013

The well-posedness of Eq. (1.1) can be guaranteed assuming that the operator B is admissible in the sense of [49, Definition 4.2.1]:

Definition 1.1. The operator $B \in \mathfrak{L}(U, X_{-1})$ is said to be an admissible control operator for $\mathbb{T}$ if for some $\tau > 0$, the operator $\mathscr{R}_\tau$ defined on $L^2(0, T; U)$ by

$$\mathscr{R}_\tau v = \int_0^\tau \mathbb{T}_{\tau-s} B v(s) \, ds$$

satisfies $\mathrm{Ran}\,\mathscr{R}_\tau \subset X$, where $\mathrm{Ran}\,\mathscr{R}_\tau$ denotes the range of the map $\mathscr{R}_\tau$.

When B is an admissible control operator for $\mathbb{T}$, system (1.1) is said to be admissible.

Of course, if B is bounded, i.e., if $B \in \mathfrak{L}(U, X)$, then B is admissible for $\mathbb{T}$. But such assumption may also hold when the operator B is not bounded, for instance when considering the wave equation controlled from its Dirichlet boundary conditions. There, the admissibility property follows from a suitable hidden regularity result for the adjoint equation of (1.1), see [36].

To be more precise, B is an admissible control operator for $\mathbb{T}$ if and only if there exist a time $T > 0$ and a constant $C_{\mathrm{ad},T}$ such that any solution of the adjoint equation

$$\varphi' = A\varphi, \quad t \in (0, T), \qquad \varphi(0) = \varphi_0 \tag{1.2}$$

with data $\varphi_0 \in \mathscr{D}(A)$ (and then in X by density) satisfies

$$\int_0^T \|B^* \varphi(t)\|_U^2 \, dt \le C_{\mathrm{ad},T}^2 \|\varphi_0\|_X^2. \tag{1.3}$$

Note that the semigroup property immediately implies that if the inequality (1.3) holds for some time T^*, it also holds for all $T > 0$.

In this work, we will always assume that B is an admissible control operator for $\mathbb{T}$. As explained in [49, Proposition 4.2.5], this implies that for every $y_0 \in X$ and $v \in L^2_{\mathrm{loc}}([0, \infty); U)$, the solution of Eq. (1.1) has a unique mild solution y which belongs to $C([0, \infty); X)$.

Let us now focus on the exact controllability property of system (1.1) in time $T^* > 0$. To be more precise, we say that system (1.1) is exactly controllable in time T^* if for all y_0 and y_f in X, there exists a control function $v \in L^2(0, T^*; U)$ such that the solution y of Eq. (1.1) satisfies $y(T^*) = y_f$.

Since we assumed that A is the generator of a strongly continuous group, using the linearity and the reversibility of Eq. (1.1), one easily checks that the exact controllability property of Eq. (1.1) in time T^* is equivalent to the a priori weaker one, the so-called null-controllability in time T^*: system (1.1) is said to be null-controllable in time T^* if for all $y_0 \in X$, there exists a control function $v \in L^2(0, T^*; U)$ such that the solution y of Eq. (1.1) satisfies $y(T^*) = 0$.

In the following, we will focus on the null-controllability property, i.e., $y_f \equiv 0$, and we shall refer to it simply as controllability.

In the sequel we assume that system (1.1) is controllable in some time T^* and we focus on the controllability property in time $T > T^*$. To be more precise, we are looking for control functions v such that the corresponding solution of Eq. (1.1) satisfies

$$y(T) = 0. \tag{1.4}$$

According to the so-called Hilbert Uniqueness Method introduced by Lions [36, 37], the controllability property is equivalent, by duality, to an observability inequality for the adjoint system (1.2) which consists in the existence of a constant C_{obs,T^*} such that for all $\varphi_0 \in X$, the solution φ of the adjoint equation (1.2) with initial data φ_0 satisfies

$$\|\varphi_0\|_X^2 \le C_{obs,T^*}^2 \int_0^{T^*} \|B^*\varphi(t)\|_U^2 \, dt. \tag{1.5}$$

Now, let $T > T^*$ and introduce δ so that $2\delta = T - T^*$ and a smooth function $\eta = \eta(t)$ such that

$$\eta \text{ smooth}, \quad \eta : \mathbb{R} \to [0,1], \quad \eta(t) = \begin{cases} 1 & \text{on } [\delta, T - \delta], \\ 0 & \text{on } \mathbb{R} \setminus (0,T). \end{cases} \tag{1.6}$$

Of course, using Eqs. (1.3), (1.5), and the fact that A is skew-adjoint, one easily checks the existence of some positive constants $C_{ad} > 0$ and $C_{obs} > 0$ such that for all initial data $\varphi_0 \in X$, the solution φ of Eq. (1.2) with initial data φ_0 satisfies

$$\int_0^T \eta(t) \|B^*\varphi(t)\|_U^2 \, dt \le C_{ad}^2 \|\varphi_0\|_X^2, \tag{1.7}$$

$$\|\varphi_0\|_X^2 \le C_{obs}^2 \int_0^T \eta(t) \|B^*\varphi(t)\|_U^2 \, dt. \tag{1.8}$$

Based on these inequalities the Hilbert Uniqueness Method yields the control of minimal norm (in $L^2((0,T), dt/\eta; U)$) by minimizing the functional

$$J(\varphi_0) = \frac{1}{2} \int_0^T \eta(t) \|B^*\varphi(t)\|_U^2 \, dt + \langle y_0, \varphi_0 \rangle_X, \tag{1.9}$$

for $\varphi_0 \in X$, where φ denotes the solution of the adjoint equation (1.2) with data φ_0.

Indeed, according to the inequalities (1.7) and (1.8), this functional J is well defined, strictly convex, and coercive on X. Therefore, it has a unique minimizer $\Phi_0 \in X$. Then, if Φ denotes the corresponding solution of Eq. (1.2) with data Φ_0, the function $V(t) = \eta(t)B^*\Phi(t)$ is a control function for Eq. (1.1). Besides, V is the control of minimal $L^2(0,T; dt/\eta; U)$-norm among all possible controls for Eq. (1.1) (i.e., so that the controlled system (1.1) fulfills the controllability requirement (1.4)).

In the sequel, we will focus on the computation of the minimizer Φ_0 of J in Eq. (1.9), which immediately gives the control function according to the formula

$$V(t) = \eta(t)B^*\Phi(t). \tag{1.10}$$

1.1.2 Contents of Chap. 1

Based on this characterization of Φ_0 as the minimizer of the functional J in Eq. (1.9), one can build an "algorithm" to approximate the minimizer in this infinite-dimensional setting. For, it suffices to apply a steepest descent or conjugate gradient iterative algorithm, for instance.

Of course, this procedure can be applied in the context of the example above in which the wave equation (1) in a bounded domain Ω with Dirichlet boundary conditions is controlled in the energy space $H_0^1(\Omega) \times L^2(\Omega)$ by means of L^2 controls localized in an open subset ω. This will be explained further in Sect. 1.7.

Once this iterative algorithm is built at the infinite-dimensional level one can mimic it for suitable numerical approximation schemes. In this way, combining the classical convergence properties of numerical schemes and the convergence properties of the iterative algorithm for the search of the minimizer of J in the functional setting above, one can get quantitative convergence results towards the control. Roughly speaking, this is the *continuous approach* to the numerical approximation of controls.

Recently, as mentioned above, a variant of this method has been developed and applied in [9] in the particular case of the wave equation. Rather than considering the HUM controls of minimal norm, characterized as the minimizers of a functional of the form J, the authors consider the control given by the classical Russell's principle, obtained as limit of an iterative process based on a stabilization property. This iterative procedure, based on the contractivity of the semigroup for exponentially decaying stabilized wave problems, applied into a numerical approximation scheme, leads to convergence rates, similar to those that the iterative methods for minimizing the functionals J as above do. Thus, the method implemented in [9] can be viewed as a particular instance of the continuous approach, see also Sect. 5.2.

The first goal of this paper is to fully develop the continuous approach in a general context of numerical approximation semigroups of the abstract evolution Eq. (1.1) based on iterative algorithms for the minimization of the functional J. Explicit convergence rates will be obtained. These results are of general application for numerous examples, including the wave equation mentioned above, see Sect. 1.7. As we shall see, these general results are similar to those stated in Theorem 1.3 obtained in [9] in the specific context of Russell's principle for the wave equation. However, the continuous approach we propose, based on the minimization of the functional J, has several advantages, and in particular the one of being applicable to non-bounded (but still admissible) control operators B and in particular in the case of boundary control for the wave equation.

The second goal of this paper is to compare these results with those one can get by means of the discrete approach which consists in controlling a finite-dimensional numerical approximation scheme of the original semigroup, in the spirit of the survey article [52] and the references therein.

To be more precise, let us consider a semi-discrete approximation of the Eq. (1.2). For all $h > 0$, we introduce the equations

$$\varphi_h' = A_h \varphi_h, \quad t \in (0,T), \quad \varphi_h(0) = \varphi_{0h}, \tag{1.11}$$

where A_h is a skew-adjoint approximation of the operator A in a finite-dimensional Hilbert space V_h embedded into X. In practice one can think of finite-difference or finite-element approximations of the PDE under consideration, for instance, h being the characteristic length of the numerical mesh.

We shall also introduce B_h^*, an approximation of the operator B^*, defined on V_h with values in some Hilbert spaces U_h.

Here, we do not give yet a precise meaning to the sense in which the sequence of operators (A_h, B_h) approximate (A, B) and converge to it as $h \to 0$. We will come back to that issue later on when stating our main results in Sect. 1.2.

Once the finite-dimensional approximation (1.11) of Eq. (1.2) has been set, one then introduces the discrete functional

$$J_h(\varphi_{0h}) = \frac{1}{2} \int_0^T \eta(t) \|B_h^* \varphi_h(t)\|_{U_h}^2 \, dt + \langle y_{0h}, \varphi_{0h} \rangle_{V_h}, \tag{1.12}$$

where φ_h is the solution of Eq. (1.11) corresponding to data $\varphi_{0h} \in V_h$ and y_{0h} is an approximation in V_h of $y_0 \in X$.

Of course, the functional J_h is a natural approximation of the continuous functional J defined by Eq. (1.9). One could then expect the minima of J_h to yield convergent approximations of the minima of the continuous functional J. It turns out that, **in general, this is not the case**. Even worse, it may even happen that, for some data y_0 to be controlled, the minimizers of these discrete functionals are not even bounded, and actually diverge exponentially as $h \to 0$, see [16, 17, 24, 52]. This is an evidence of the lack of Γ-convergence of the functionals J_h towards J.

This instability is due to spurious high-frequency numerical components that make the discrete versions of the observability inequalities to blow up as $h \to 0$, see, e.g., [38, 48, 50].

However, once we have understood that these instabilities arise at high frequencies, one can develop filtering techniques which consist, essentially, in restricting the functionals J_h to subspaces of V_h in which they are uniformly coercive and so that these subspaces, as $h \to 0$, cover the whole space X, thus ensuring the Γ-convergence of the restricted functionals. These subspaces can be chosen in various manners: we refer to [12, 28, 41, 51] for Fourier filtering techniques, [42] for bi-grid methods, [43] for wavelet approximations, and [6, 7, 10, 13] for other discretization methods designed to attenuate these high-frequency pathologies. In this way one can obtain the convergence of discrete controls towards the continuous one and even convergence rates, based on the results in [15], see [16, 17].

But, it is important to note that the minimizers obtained by minimizing the functionals J_h on strict subspaces of V_h do not yield exact controls of the finite-dimensional dynamics but rather partial controls, in which the controllability requirement at time $t = T$ is relaxed so that a suitable projection of the solution is controlled. In other words, relaxing the minimization process to a subspace of the whole space V_h yields a relaxation of the control requirement as well.

This discrete analysis is based on a deep understanding of the finite-dimensional dynamics of Eq. (1.11) in contrast with the continuous approach that uses simply the control results for the continuous system and the classical results on the convergence of finite-dimensional approximations.

The third and last goal of this chapter is to compare the convergence results obtained by the continuous approach with those one gets applying the discrete one. As we shall see, finally, the filtering methods developed in the discrete setting can also be understood in the continuous context, as an efficient projection of the numerical approximation of the gradient-like iteration procedures developed in the continuous frame.

Our main results end up unifying, to a large extent, both the continuous and the discrete approaches.

1.2 Main Results

1.2.1 An "Algorithm" in an Infinite-Dimensional Setting

In the abstract setting of the previous section, let us introduce the so-called Gramian operator Λ_T defined on X by

$$\forall (\varphi_0, \psi_0) \in X^2, \quad \langle \Lambda_T \varphi_0, \psi_0 \rangle_X = \int_0^T \eta(t) \langle B^* \varphi(t), B^* \psi(t) \rangle_U \, dt, \qquad (1.13)$$

where $\varphi(t), \psi(t)$ are the corresponding solutions of Eq. (1.2).

Obviously, this Gramian operator is nothing but the gradient of the quadratic term entering in the functional J and therefore plays a key role when identifying the Euler–Lagrange equations associated to the minimization of J and when building gradient-like iterative algorithms. In particular, $\Phi_0 \in X$ is a critical point of J (hence automatically a minimum since J is strictly convex) if and only if

$$\Lambda_T \Phi_0 + y_0 = 0. \qquad (1.14)$$

Note that this Gramian operator can be written, at least formally, as

$$\Lambda_T = \int_0^T \eta(t) e^{-tA} BB^* e^{tA} \, dt.$$

Under this form, one immediately sees that Λ_T is a self-adjoint nonnegative operator, and that it is bounded and positive definite when Eqs. (1.7) and (1.8) hold.

Of course, estimates (1.7) and (1.8), which guarantee that J is well defined, coercive, and strictly convex, and hence the uniqueness of the minimizer to J, also imply the existence and uniqueness of a solution $\Phi_0 \in X$ of Eq. (1.14).

Before going further, let us explain that, when assuming Eqs. (1.7) and (1.8), if $s \geq 0$, for $y_0 \in X_s$ the solution Φ_0 of Eq. (1.14) also belongs to X_s and there exists a constant C_s such that

$$\|\Phi_0\|_s \leq C_s \|y_0\|_s. \tag{1.15}$$

This is a consequence of the regularity results derived in [15] obtained for abstract conservative systems in which the fact of having introduced the time cutoff function η in Eq. (1.6) within the definition of the Gramian Λ_T plays a critical role. Otherwise, if $\eta \equiv 1$ our analysis would have to be restricted to bounded control operators such that BB^* maps X_p to X_p for each $p \in [0, \lceil s \rceil]$, see [15].

Note that the results in [15] can also be seen as an abstract counterpart of the results in [11], which state that, in the case of the wave equation with distributed controls (hence corresponding to the case of bounded control operators), the Gramian with this cutoff function $\eta = \eta(t)$ in time and a control operator $BB^* \in \cap_{p>0} \mathcal{L}(X_p)$ maps X_s to X_s for all $s \geq 0$. The results in [11] are even more precise when working on a compact manifold without boundary, in which case it is proved that the inverse of the Gramian is a pseudo-differential operator that preserves the regularity of the data.

To fully develop the continuous approach to the numerical approximation of the controls, we implement the steepest descent algorithm for the minimization of the functional J in Eq. (1.9). But for doing that it is more convenient to have an alternate representation of the Gramian.

Let $\varphi_0 \in X$ and φ be the corresponding solution of Eq. (1.2). Then solve

$$\psi' = A\psi - \eta BB^*\varphi, \quad t \in (0,T), \quad \psi(T) = 0. \tag{1.16}$$

Then, as it can be easily seen,

$$\Lambda_T \varphi_0 = \psi(0),$$

where ψ solves Eq. (1.16) and φ is the solution of Eq. (1.2).

The steepest descent algorithm then reads as follows:

- *Initialization*: Define

$$\varphi_0^0 = 0. \tag{1.17}$$

- *Iteration*: For $\varphi_0^k \in X$, define φ_0^{k+1} by

$$\varphi_0^{k+1} = \varphi_0^k - \rho(\Lambda_T \varphi_0^k + y_0), \tag{1.18}$$

where $\rho > 0$ is a fixed parameter, whose (small enough) value will be specified later on.

We shall then show the following results:

Theorem 1.1. *Let $s \geq 0$. Assume that the estimates (1.7) and (1.8) hold true. Let $y_0 \in X_s$ and $\Phi_0 \in X$ be the solution of Eq. (1.14).*

Then setting $\rho_0 > 0$ as

$$\rho_0 = \frac{2}{C_{ad}^4 C_{obs}^2}, \tag{1.19}$$

for all $\rho \in (0, \rho_0)$, the sequence φ_0^k defined by Eqs. (1.17) and (1.18) satisfies, for some constant $\delta \in (0, 1)$ given by

$$\delta(\rho) := \sqrt{1 - 2\frac{\rho}{C_{obs}^2} + \rho^2 C_{ad}^4}, \tag{1.20}$$

that for all $k \in \mathbb{N}$,

$$\left\| \varphi_0^k - \Phi_0 \right\|_X \leq C \delta^k \|y_0\|_X. \tag{1.21}$$

Besides, $\Phi_0 \in X_s$ and for all $k \in \mathbb{N}$, the sequence φ_0^k belongs to X_s. The sequence φ_0^k also strongly converges to Φ_0 in X_s and satisfies, for some constant C_s independent of $\Phi_0 \in X_s$ and $k \in \mathbb{N}$:

$$\left\| \varphi_0^k - \Phi_0 \right\|_s \leq C_s (1 + k^s) \delta^k \|y_0\|_s, \quad k \in \mathbb{N}. \tag{1.22}$$

The first statement (1.21) in Theorem 1.1 is nothing but the application of the well-known results on the convergence rate for the steepest descent method when minimizing quadratic coercive and continuous functionals in Hilbert spaces [8]. However, the result (1.22) is new and relies in an essential manner on the fact that the Gramian operator preserves the regularity properties of the data to be controlled, a fact that was proved in [15] and for which the weight function $\eta = \eta(t)$ plays a key role.

Also note that the results in Theorem 1.1 are written in terms of the norms of y_0, but we will rather prove the following stronger results (according to Eq. (1.15)):

$$\left\| \varphi_0^k - \Phi_0 \right\|_X \leq \delta^k \|\Phi_0\|_X, \tag{1.23}$$

and, if $y_0 \in X_s$,

$$\left\| \varphi_0^k - \Phi_0 \right\|_s \leq C_s (1 + k^s) \delta^k \|\Phi_0\|_s, \quad k \in \mathbb{N}. \tag{1.24}$$

Of course, these convergence results also imply that the sequence $v^k = \eta B^* \varphi^k$, where φ^k is the solution of Eq. (1.2) with initial data φ_0^k, converge to the control V given by Eq. (1.10):

$$\left\| v^k - V \right\|_{L^2(0,T;dt/\eta;U)} \leq C \delta^k \|y_0\|_X. \tag{1.25}$$

Note that, in general, Eq. (1.22) also gives estimates on the convergence of v^k towards V in stronger norms when the data y_0 to be controlled lies in X_s for some $s \geq 0$.

1.2.2 The Continuous Approach

Following the "algorithm" developed in Theorem 1.1, we now approximate the sequence φ_0^k constructed in Eqs. (1.17) and (1.18). A way of doing that is to introduce operators A_h and B_h as above and to define the discrete operator

$$\Lambda_{Th} = \int_0^T \eta(t) e^{-tA_h} B_h B_h^* e^{tA_h} \, dt.$$

To be more precise, we shall assume that we have an extension map $E_h : V_h \to X$ that induces an Hilbert structure on V_h endowed by the norm $\|\cdot\|_h = \|E_h \cdot\|_X$. We further assume that, for each $h > 0$, A_h is skew-adjoint with respect to that scalar product, so that Λ_{Th} is self-adjoint in V_h.

Classically, for the numerical method to be consistent, it is assumed that for smooth initial data $\varphi \in \cap_{s>0} X_s$, $(E_h R_h - Id)\varphi$ strongly converge to zero in X as $h \to 0$, where R_h is a restriction operator from X to V_h. But for our purpose, we need a slightly different version of it (though of course they are related), see Assumption 2 in Eq. (1.29) below.

Here, we shall rather assume the two following conditions:

Assumption 1. There exist $s > 0$, $\theta > 0$ and $C > 0$ so that for all $h > 0$

$$\|E_h R_h \varphi_0\|_X \leq C \|\varphi_0\|_X, \quad \varphi \in X, \tag{1.26}$$

$$\|(E_h R_h - Id_X)\varphi_0\|_X \leq C h^\theta \|\varphi_0\|_s, \quad \varphi \in X_s, \tag{1.27}$$

$$\|E_h \Lambda_{Th} R_h \varphi_0\|_X \leq C \|\varphi_0\|_X, \quad \varphi \in X, \tag{1.28}$$

$$\|(E_h \Lambda_{Th} R_h - E_h R_h \Lambda_T)\varphi_0\|_X \leq C h^\theta \|\varphi_0\|_s, \quad \varphi \in X_s. \tag{1.29}$$

Assumption 2. The norms of the operators Λ_{Th} in $\mathfrak{L}(V_h)$ are uniformly bounded with respect to $h > 0$:

$$\mathscr{C}_{ad}^2 := \sup_{h \geq 0} \|\Lambda_{Th}\|_{\mathfrak{L}(V_h)} < \infty, \tag{1.30}$$

where, when $h = 0$, we use the notation $V_0 = X$ and $\Lambda_{T0} = \Lambda_T$.

Before going further, let us emphasize that Assumption 2, though straightforward when the observation operators are uniformly bounded with respect to the $\mathfrak{L}(X_h, U_h)$ norms, is not obvious when dealing with boundary controls, for instance. Indeed, in that case, one should be careful and prove a uniform admissibility result (here and in the following, "uniform" always refers to the dependence on the discretization parameter(s)). Also note that Assumption 2 together with Eq. (1.26) implies Eq. (1.28).

We now have the following result:

Theorem 1.2. *Assume that Assumptions 1 and 2 hold. Define ρ_1 by*

$$\rho_1 = \min\{\rho_0, 2/\mathscr{C}_{ad}^2\}, \tag{1.31}$$

where ρ_0 is given by Theorem 1.1.

Let $\rho \in (0, \rho_1)$. Let $y_0 \in X_s$ and $(y_{0h})_{h>0}$ be a sequence of functions such that for all $h > 0$, $y_{0h} \in V_h$.

For each $h > 0$, define the sequence φ_{0h}^k by induction, inspired in the statement of Theorem 1.1, as follows:

$$\varphi_{0h}^0 = 0, \quad \forall k \in \mathbb{N}, \quad \varphi_{0h}^{k+1} = \varphi_{0h}^k - \rho \left(\Lambda_{Th} \varphi_{0h}^k + y_{0h} \right). \tag{1.32}$$

Then consider the sequence φ_0^k defined by induction by Eqs. (1.17) and (1.18) with this same parameter ρ.

Then there exists a constant $C > 0$ independent of $h > 0$ such that for all $k \in \mathbb{N}$,

$$\left\| E_h \varphi_{0h}^k - \varphi_0^k \right\|_X \leq k\rho \left\| E_h y_{0h} - y_0 \right\|_X + Ckh^\theta \left\| y_0 \right\|_s . \tag{1.33}$$

Then, using Theorems 1.1 and 1.2 together, we get the following convergence theorem.

Theorem 1.3. *Assume that Assumptions 1 and 2 hold.*

Let $y_0 \in X_s$ and $\rho \in (0, \rho_1)$, ρ_1 given by Eq. (1.31). Let $(y_{0h})_{h>0}$ be a sequence such that for all $h > 0$,

$$\left\| E_h y_{0h} - y_0 \right\|_X \leq Ch^\theta \left\| y_0 \right\|_s . \tag{1.34}$$

Then, for all $h > 0$, setting

$$K_h^c = \left\lfloor \theta \frac{\log(h)}{\log(\delta)} \right\rfloor , \tag{1.35}$$

where δ is given by Eq. (1.20), we have, for some constant C independent of h,

$$\left\| E_h \varphi_{0h}^{K_h^c} - \Phi_0 \right\|_X \leq C |\log(h)|^{\max\{1,s\}} h^\theta \left\| y_0 \right\|_s , \tag{1.36}$$

where $\varphi_{0h}^{K_h^c}$ is the K_h^c-iterate of the sequence φ_{0h}^k defined by Eq. (1.32).

This is the so-called *continuous approach* for building numerical approximations of the controls.

At this level it is convenient to underline a number of issues:

- The approximate controls we obtain in this way do not control the discrete dynamics or some of its projections. They are simply obtained as approximations of the continuous control by mimicking at the discrete level the iterative algorithm of Theorem 1.1.
- The result above holds provided the number of iterations is limited by the threshold given by Eq. (1.35). Indeed, in case the iterative algorithm would be continued after this step, the error estimate would deteriorate as the numerical experiments show; see Sect. 1.7.

As mentioned above, the algorithm above and the error estimates we obtain are similar to those in [9] where the iterative process proposed by Russell to obtain controllability out of stabilization results is mimicked at the discrete level. The number of iterations in [9] is of the order of $\lfloor \theta | \log(h) | m \rfloor$, where m is a constant that enters in the continuous stabilization property of the dissipative operator $A - BB^*$, and the error obtained that way is $h^\theta | \log(h)|^2$. But the results in [9] apply only in the context of bounded control operators and they do not yield the control of minimal L^2-norm, whereas our approach applies under the weaker admissibility assumption on the control operator and yields effective approximations of the minimal norm controls (suitably weighted in time).

Note that estimates (1.36) also imply that the sequence $v_h^k = \eta B_h^* \varphi_h^k$, defined for $k \geq 0$ with $\varphi_h^k(t) = \exp(tA_h)\varphi_{0h}^k$ satisfies that $v_h^{K_h^c}$ is close to V in Eq. (1.10) with some bounds (usually the same) on the error term. We do not state precisely the corresponding results since it would require to introduce further assumptions on the way the spaces U_h approximate U.

1.2.3 The Discrete Approach

As we have mentioned above, the discrete approach is based on the analysis and use of the controllability properties of the approximated discrete dynamics to build efficient numerical approximations of the controls.

The main difference when implementing it is that it requires the following uniform coercivity assumption on the Gramian operator:

Assumption 3. There exists a constant $\mathscr{C}$ such that for all $h \geq 0$ and $\varphi_{0h} \in V_h$,

$$\|\varphi_{0h}\|_h^2 \leq \mathscr{C}^2 \langle \Lambda_{Th}\varphi_{0h}, \varphi_{0h} \rangle_h, \tag{1.37}$$

where, for $h = 0$, we use the notation $V_0 = X$ and $\Lambda_{T0} = \Lambda_T$.

Note that Assumption 3 states the uniform coercivity of the operators Λ_{Th}, or equivalently, the uniform observability for the approximated semigroups. This assumption often fails and is only guaranteed to hold in suitable subspaces of V_h, after applying suitable filtering mechanisms (see [16, 23, 24]). Indeed, the classical numerical methods employed to approximate Λ_T by Λ_{Th} that are usually based on replacing the wave equation by a numerical approximation counterpart usually provide discrete operators Λ_{Th} that violate this uniform observability assumption. Hence providing a subspace V_h satisfying Eq. (1.37) requires a careful analysis of the observability properties of the discrete dynamics, a fact that is not necessary when developing the continuous approach.

In any case, under Assumption 3, we can prove the following stronger version of Theorem 1.2:

Theorem 1.4. *Assume that Assumptions 1, 2, and 3 hold. Define*

$$\rho_2 = \frac{2}{\mathscr{C}_{\mathrm{ad}}^4 \mathscr{C}^2} \tag{1.38}$$

and consider $\rho \in (0, \rho_2)$. Let $y_0 \in \mathscr{D}(A^s)$ and $(y_{0h})_{h>0}$ be a sequence of functions such that for all $h > 0$, $y_{0h} \in V_h$.

For each $h > 0$, define the sequence φ_{0h}^k by induction as in Eq. (1.32). Then consider the sequence φ^k defined by induction by Eqs. (1.17) and (1.18) with this same parameter ρ.

Then there exists a constant $C > 0$ independent of $h > 0$ such that for all $k \in \mathbb{N}$,

$$\left\| E_h \varphi_{0h}^k - \varphi_0^k \right\|_X \le C \left(\|E_h y_{0h} - y_0\|_X + h^\theta \|y_0\|_s \right). \tag{1.39}$$

Then, using Theorems 1.1 and 1.4 together, we get the following counterpart of Theorem 1.3:

Theorem 1.5. *Let us suppose that Assumptions 1, 2, and 3 hold.*
Let $y_0 \in X_s$ and $\rho \in (0, \rho_2)$, ρ_2 given by Eq. (1.38).
Let $(y_{0h})_{h>0}$ be a sequence such that Eq. (1.34) holds.
Then, for all $h > 0$, setting

$$K_h^d = \left\lfloor \theta \frac{\log(h)}{\log(\delta)} - (s+1) \frac{\log(|\log(h)|)}{\log(\delta)} \right\rfloor, \tag{1.40}$$

where δ is given by Eq. (1.20), we have, for some constant C independent of h and k,

$$\left\| E_h \varphi_{0h}^k - \Phi_0 \right\|_X \le C h^\theta \|y_0\|_s, \quad k \ge K_h^d, \tag{1.41}$$

where φ_{0h}^k is the k-iterate of the sequence φ_{0h}^k defined by Eq. (1.32).

Note that, under Assumptions 1, 2, and 3, for all $y_{0h} \in V_h$, the equation

$$\Lambda_{Th} \Phi_{0h} + y_{0h} = 0 \tag{1.42}$$

has a unique solution Φ_{0h}, on which we have a uniform bound:

$$\|\Phi_{0h}\|_h \le \mathscr{C}^2 \|y_{0h}\|_h, \tag{1.43}$$

where $\mathscr{C}$ is the constant in Assumption 3.

Now, since k can be made arbitrarily large in Theorem 1.5, if $y_0 \in X_s$ and y_{0h} denotes an approximation of y_0 that satisfies Eq. (1.34), setting Φ_{0h} the solution of Eq. (1.42), we have

$$\|E_h \Phi_{0h} - \Phi_0\|_X \le C h^\theta \|y_0\|_s, \tag{1.44}$$

where Φ_0 is the solution of Eq. (1.14). Indeed, in that case, it is very easy to check that at $h > 0$ fixed, as $k \to \infty$, the sequence ϕ_{0h}^k converges to $\Phi_{0,h}$ given by Eq. (1.42) in V_h, see, e.g., Theorem 1.6.

This is the convergence result obtained in [16], using another proof, directly based on the smoothness of the trajectory of the minimizer Φ_0 when $y_0 \in X_s$. We refer to [16] for numerical evidences on the fact that the convergence rates (1.44) are close to sharp. We will also illustrate this fact in Sect. 1.7.

1.2.4 Outline of Chap. 1

Chapter 1 is organized as follows. In Sect. 1.3 we prove Theorem 1.1. In Sect. 1.4 we give the proofs of Theorems 1.2 and 1.3. In Sect. 1.5 we prove Theorems 1.4 and 1.5. We shall then compare the two approaches in Sect. 1.6. In Sect. 1.7 we present some applications of these abstract results, in particular to the wave equation. In Sect. 1.8 we show that some data assimilation problems can be treated by the methods developed in this book.

1.3 Proof of the Main Result on the Continuous Setting

This section is devoted to the proof of Theorem 1.1. We shall then fix $T > 0$ so that estimates (1.7) and (1.8) hold. Given $y_0 \in X_s$, $\Phi_0 \in X$ is chosen to be the unique solution of Eq. (1.14).

Let then φ_0^k be sequence defined by the induction formulae (1.17) and (1.18).

1.3.1 Classical Convergence Results

First we prove Eq. (1.23) which is classical and corresponds to the usual proof of convergence of the steepest descent algorithm for quadratic convex functionals. We provide it only for completeness and later use.

Proof (Proof of estimate (1.23)). Using Eq. (1.18), and subtracting to it Φ_0, we get

$$\varphi_0^{k+1} - \Phi_0 = \varphi_0^k - \Phi_0 - \rho(\Lambda_T \varphi_0^k + y_0) = \varphi_0^k - \Phi_0 - \rho\Lambda_T(\varphi_0^k - \Phi_0), \qquad (1.45)$$

where the last identity follows from the definition of Φ_0 in Eq. (1.14).

But, for any $\psi \in X$,

$$\|(I - \rho\Lambda_T)\psi\|_X^2 = \|\psi\|_X^2 - 2\rho\langle\Lambda_T\psi, \psi\rangle_X + \rho^2\|\Lambda_T\psi\|_X^2.$$

Hence, using that Eqs. (1.7) and (1.8) can be rewritten as

$$\frac{1}{C_{\text{obs}}^2}\|\psi\|_X^2 \leq \left\|\Lambda_T^{1/2}\psi\right\|_X^2 \leq C_{\text{ad}}^2\|\psi\|_X^2,$$

we get that

$$\|(I-\rho\Lambda_T)\psi\|_X^2 \leq \left(1 - 2\frac{\rho}{C_{\text{obs}}^2} + \rho^2 C_{\text{ad}}^4\right)\|\psi\|_X^2. \tag{1.46}$$

Note that, according to Eqs. (1.7) and (1.8), $C_{\text{obs}}^2 C_{\text{ad}}^2 \geq 1$ and thus for all $\rho > 0$, the quadratic form $1 - 2\rho/C_{\text{obs}}^2 + \rho^2 C_{\text{ad}}^4$ is nonnegative.

Thus, for any $\rho > 0$ such that $\rho \in (0,\rho_0)$ with ρ_0 as in Eq. (1.19), and setting $\delta(\rho)$ as in Eq. (1.20), $\delta(\rho)$ belongs to $(0,1)$ and

$$\|(I-\rho\Lambda_T)\|_{\mathcal{L}(X)} \leq \delta(\rho). \tag{1.47}$$

From Eq. (1.45), we obtain

$$\left\|\varphi_0^{k+1} - \Phi_0\right\|_X \leq \delta\left\|\varphi_0^k - \Phi_0\right\|_X. \tag{1.48}$$

Of course, Eq. (1.48) immediately implies Eq. (1.23). $\square$

1.3.2 Convergence Rates in X_s

Here, our goal is to show the convergence of the sequence φ_0^k constructed in Eqs. (1.17) and (1.18) in the space X_s.

Proof (Proof of the convergence in X_s). When $s \in \mathbb{R}_+$, the convergence estimate (1.24) is deduced by interpolation between the results obtained for $\lfloor s \rfloor$ and $\lceil s \rceil$. Hence, in the following, we focus on the proof of Eq. (1.24) for integers $s \in \mathbb{N}$. Besides, the case $s = 0$ is already done in Eq. (1.23) so we will be interested in the case $s \in \mathbb{N}$ and $s \geq 1$.

Step 1. The Gramian operator maps $\mathscr{D}(A^s)$ to $\mathscr{D}(A^s)$. For $\psi_0 \in X$, introduce the function $\Psi_0 \in D(A)$ defined by $A\Psi_0 = \psi_0$. Then the solutions Ψ and ψ of Eq. (1.2) with corresponding initial data Ψ_0 and ψ_0 satisfy, for all $t \in (0,T)$, $\Psi'(t) = A\Psi(t) = \psi(t)$.

Hence, if $\varphi_0 \in \mathscr{D}(A)$ and $\psi_0 \in X$, denoting by φ the solution of Eq. (1.2) with data φ_0,

$$\langle\Lambda_T\varphi_0, \psi_0\rangle_X = \int_0^T \eta(t)\langle B^*\varphi(t), B^*\Psi'(t)\rangle_U \, dt$$

$$= -\int_0^T \eta(t)\langle B^*\varphi'(t), B^*\Psi(t)\rangle_U \, dt - \int_0^T \eta'(t)\langle B^*\varphi(t), B^*\Psi(t)\rangle_U \, dt.$$

Of course, using Eqs. (1.7) and (1.3), this implies that

$$|\langle \Lambda_T \varphi_0, \psi_0 \rangle_X| \leq C_{ad}^2 \|A\varphi_0\|_X \|\Psi_0\|_X + \|\eta'\|_{L^\infty} C_{ad,T}^2 \|\varphi_0\|_X \|\Psi_0\|_X$$
$$\leq \|A\varphi_0\|_X \|A^{-1}\psi_0\|_X \left(C_{ad}^2 + C_{ad,T}^2 \|\eta'\|_{L^\infty} \|A^{-1}\|_{\mathcal{L}(X)} \right).$$

Therefore, Λ_T maps $\mathscr{D}(A)$ to itself.

Of course, the case of an integer $s \in \mathbb{N}$ strictly larger than 1 can be treated similarly and is left to the reader. Then, by interpolation, this also implies that for all $s \geq 0$, Λ_T maps X_s to X_s.

This step already indicates that for each $k \in \mathbb{N}$, φ_0^k constructed by the induction formulae (1.17) and (1.18) belongs to X_s provided that $y_0 \in X_s$.

Step 2. First estimate on the commutator $[\Lambda_T, A]$. Take φ_0 and ψ_0 in $\mathscr{D}(A)$. From the previous step, we know that $[\Lambda_T, A]\varphi_0 \in X$, and we can then take its scalar product with $\psi_0 \in \mathscr{D}(A)$:

$$\langle [\Lambda_T, A]\varphi_0, \psi_0 \rangle_X = \langle \Lambda_T A\varphi_0, \psi_0 \rangle_X - \langle A\Lambda_T \varphi_0, \psi_0 \rangle_X$$
$$= \langle \Lambda_T A\varphi_0, \psi_0 \rangle_X + \langle \Lambda_T \varphi_0, A\psi_0 \rangle_X$$
$$= \int_0^T \eta(t)\langle B^*\varphi'(t), B^*\psi(t)\rangle_U \, dt + \int_0^T \eta \langle B^*\varphi(t), B^*\psi'(t)\rangle_U \, dt$$
$$= -\int_0^T \eta'(t)\langle B^*\varphi(t), B^*\psi(t)\rangle_U \, dt,$$

where φ and ψ are the solutions of Eq. (1.2) with data φ_0 and ψ_0, respectively. Hence, using Eq. (1.3), we obtain

$$|\langle [\Lambda_T, A]\varphi_0, \psi_0 \rangle_X| \leq C_{ad,T}^2 \|\eta'\|_{L^\infty} \|\varphi_0\|_X \|\psi_0\|_X, \tag{1.49}$$

and the operator $[\Lambda_T, A]$ can be extended as a continuous operator from X to X and

$$\|[\Lambda_T, A]\|_{\mathcal{L}(X)} \leq C_{ad,T}^2 \|\eta'\|_{L^\infty}. \tag{1.50}$$

Step 3. Convergence in $\mathscr{D}(A)$. Apply A in the identity (1.45):

$$A\left(\varphi_0^{k+1} - \Phi_0\right) = A\left(\varphi_0^k - \Phi_0\right) - \rho A\Lambda_T(\varphi_0^k - \Phi_0)$$
$$= A\left(\varphi_0^k - \Phi_0\right) - \rho \Lambda_T A(\varphi_0^k - \Phi_0) + \rho[\Lambda_T, A](\varphi_0^k - \Phi_0). \tag{1.51}$$

Then, using Eqs. (1.47) and (1.50), we obtain

$$\left\|A\left(\varphi_0^{k+1} - \Phi_0\right)\right\|_X \leq \delta \left\|A\left(\varphi_0^k - \Phi_0\right)\right\|_X + \rho \|[\Lambda_T, A]\|_{\mathcal{L}(X)} \left\|\left(\varphi_0^k - \Phi_0\right)\right\|_X, \tag{1.52}$$

and, using Eq. (1.23),

$$\left\|A\left(\varphi_0^{k+1} - \Phi_0\right)\right\|_X \leq \delta \left\|A\left(\varphi_0^k - \Phi_0\right)\right\|_X + \rho \delta^k \|[\Lambda_T, A]\|_{\mathcal{L}(X)} \|\Phi_0\|_X.$$

Therefore,

$$\frac{1}{\delta^{k+1}}\left\|A\left(\varphi_0^{k+1}-\Phi_0\right)\right\|_X - \frac{1}{\delta^k}\left\|A\left(\varphi_0^k-\Phi_0\right)\right\|_X \leq \frac{\rho}{\delta}\left\|[\Lambda_T,A]\right\|_{\mathcal{L}(X)}\left\|\Phi_0\right\|_X.$$

Summing up these inequalities, we obtain, for all $k \in \mathbb{N}$,

$$\left\|A\left(\varphi_0^k-\Phi_0\right)\right\|_X \leq \delta^k\left(\left\|A\Phi_0\right\|_X + \frac{\rho k}{\delta}\left\|[\Lambda_T,A]\right\|_{\mathcal{L}(X)}\left\|\Phi_0\right\|_X\right). \tag{1.53}$$

Step 4. Higher-order estimates (1.24). They are left to the reader as they are very similar to those obtained in Eq. (1.53). They are obtained by induction for $s \in \mathbb{N}$.

The idea is to write

$$A^s\left(\varphi_0^{k+1}-\Phi_0\right) = (I-\rho\Lambda_T)A^s\left(\varphi_0^k-\Phi_0\right) + \rho[\Lambda_T,A^s]\left(\varphi_0^k-\Phi_0\right),$$

and use the fact that $[\Lambda_T,A^s]$ is bounded as an operator from $\mathscr{D}(A^{s-1})$ to X, which can be proved similarly as in *Step 2*. Then one easily gets that

$$\left\|A^s\left(\varphi_0^{k+1}-\Phi_0\right)\right\|_X \leq \delta\left\|A^s\left(\varphi_0^k-\Phi_0\right)\right\|_X$$
$$+ \rho\left\|[\Lambda_T,A^s]\right\|_{\mathcal{L}(\mathscr{D}(A^{s-1}),X)}\left\|A^{s-1}\left(\varphi_0^{k+1}-\Phi_0\right)\right\|_X.$$

An easy induction argument then yields Eq. (1.24) for all $s \in \mathbb{N}$.

To conclude Eq. (1.24) for $s \geq 0$, we interpolate Eq. (1.24) between the two consecutive integers $\lfloor s \rfloor$ and $\lceil s \rceil$. $\square$

Remark 1.1. Note that, actually, the smoothness $\eta \in C^\infty(\mathbb{R})$ is not really needed to get Theorem 1.1. The assumption $\eta \in C^{\lceil s \rceil}(\mathbb{R})$ would be enough.

Also note that, when BB^* maps $\mathscr{D}(A^p)$ to $\mathscr{D}(A^p)$ for all $p \in \mathbb{N}$, one can even choose η as being the step function $\eta(t) = 1$ on $(0,T^*)$ (where T^* is such that the observability estimate (1.5) holds) and vanishing outside $(0,T^*)$.

These remarks are of course related to the fact that in these two cases, the needed integrations by parts run smoothly, similarly as in [15].

1.4 The Continuous Approach

In this section, we suppose that Assumptions 1 and 2 hold.

1.4.1 Proof of Theorem 1.2

Proof (Theorem 1.2). In the following, we use the notations introduced in Theorem 1.2. All the constants that will appear in the proof below, denoted by a generic C that may change from line to line, are independent of $h > 0$ and $k \in \mathbb{N}$.

Subtracting Eq. (1.18) to Eq. (1.32), we obtain

$$\varphi_{0h}^{k+1} - R_h \varphi_0^{k+1} = \varphi_{0h}^k - R_h \varphi_0^k - \rho \left(y_{0h} - R_h y_0 \right) - \rho \left(\Lambda_{Th} \varphi_{0h}^k - R_h \Lambda_T \varphi_0^k \right)$$

$$= (I - \rho \Lambda_{Th}) \left(\varphi_{0h}^k - R_h \varphi_0^k \right) - \rho \left(y_{0h} - R_h y_0 \right) + \rho \left(R_h \Lambda_T - \Lambda_{Th} R_h \right) \varphi_0^k.$$

Hence,

$$\left\| \varphi_{0h}^{k+1} - R_h \varphi_0^{k+1} \right\|_h \leq \left\| (I - \rho \Lambda_{Th}) \left(\varphi_{0h}^k - R_h \varphi_0^k \right) \right\|_h$$
$$+ \rho \left\| y_{0h} - R_h y_0 \right\|_h + \rho \left\| (R_h \Lambda_T - \Lambda_{Th} R_h) \varphi_0^k \right\|_h. \quad (1.54)$$

But, for $\varphi_h \in V_h$,

$$\left\| (I - \rho \Lambda_{Th}) \varphi_h \right\|_h^2 = \left\| \varphi_h \right\|_h^2 - 2\rho \left\| \Lambda_{Th}^{1/2} \varphi_h \right\|_h^2 + \rho^2 \left\| \Lambda_{Th} \varphi_h \right\|_h^2$$

$$= \left\| \varphi_h \right\|_h^2 - 2\rho \left\| \Lambda_{Th}^{1/2} \varphi_h \right\|_h^2 + \rho^2 \left\| \Lambda_{Th} \varphi_h \right\|_h^2$$

$$\leq \left\| \varphi_h \right\|_X^2 - 2\rho \left\| \Lambda_{Th}^{1/2} \varphi_h \right\|_h^2 + \rho^2 \left\| \Lambda_{Th}^{1/2} \right\|_{\mathfrak{L}(V_h)}^2 \left\| \Lambda_{Th}^{1/2} \varphi_h \right\|_h^2.$$

Hence, if we impose $\rho \in (0, \rho_1)$, where $\rho_1 = \min\{\rho_0, 2/\mathscr{C}_{ad}^2\}$ as in Eq. (1.31) (with $\mathscr{C}_{ad}$ given by Assumption 2),

$$-2 + \rho \left\| \Lambda_{Th}^{1/2} \right\|_{\mathfrak{L}(V_h)}^2 \leq -2 + \rho \mathscr{C}_{ad}^2 \leq 0,$$

and then for all $\varphi_h \in V_h$,

$$\left\| (I - \rho \Lambda_{Th}) \varphi_h \right\|_h^2 \leq \left\| \varphi_h \right\|_h^2. \quad (1.55)$$

Accordingly, for $\rho \in (0, \rho_1)$,

$$\left\| (I - \rho \Lambda_{Th}) \left(\varphi_{0h}^k - R_h \varphi_0^k \right) \right\|_h \leq \left\| \left(\varphi_{0h}^k - R_h \varphi_0^k \right) \right\|_h. \quad (1.56)$$

Equation (1.29) in Assumption 1 also yields

$$\left\| (R_h \Lambda_T - \Lambda_{Th} R_h) \varphi_0^k \right\|_h \leq C h^\theta \left\| \varphi_0^k \right\|_s. \quad (1.57)$$

Using the fact that, according to estimates (1.15), (1.24), there is a constant C independent of k and $h > 0$ such that for all $k \in \mathbb{N}$,

$$\left\| \varphi_0^k \right\|_s \le C \left\| y_0 \right\|_s, \tag{1.58}$$

we derive

$$\left\| (R_h \Lambda_T - \Lambda_{Th} R_h) \varphi_0^k \right\|_h \le C h^\theta \left\| y_0 \right\|_s. \tag{1.59}$$

Thus, using Eqs. (1.54), (1.56), and (1.59), we obtain

$$\left\| \varphi_{0h}^{k+1} - R_h \varphi_0^{k+1} \right\|_h \le \left\| \varphi_{0h}^k - R_h \varphi_0^k \right\|_h + \rho \left\| y_{0h} - R_h y_0 \right\|_h + C \rho h^\theta \left\| y_0 \right\|_s, \tag{1.60}$$

where C is a constant independent of k and $h > 0$.

Summing up Eq. (1.60), we obtain

$$\left\| E_h \varphi_{0h}^k - E_h R_h \varphi_0^k \right\|_X = \left\| \varphi_{0h}^k - R_h \varphi_0^k \right\|_h \le k \rho \left\| y_{0h} - R_h y_0 \right\|_h + C k \rho h^\theta \left\| y_0 \right\|_s.$$

Finally, according to Eq. (1.58), estimate (1.27) yields

$$\left\| (E_h R_h - Id_X) \varphi_0^k \right\|_X \le C h^\theta \left\| y_0 \right\|_s, \tag{1.61}$$

and thus Eq. (1.33) follows immediately. $\square$

1.4.2 Proof of Theorem 1.3

Proof (Theorem 1.3). Using Theorems 1.1 and 1.2 and the estimate (1.34), we obtain that, for some constant $C > 0$ independent of h,

$$\left\| E_h \varphi_{0h}^k - \Phi_0 \right\|_X \le C \left\| y_0 \right\|_s \left((1 + k^s) \delta^k + k \rho h^\theta \right). \tag{1.62}$$

We then optimize the right-hand side of this estimate in k, thus yielding approximately K_h^c as in Eq. (1.35). Estimate (1.36) immediately follows from the definition of K_h^c. $\square$

1.5 Improved Convergence Rates: The Discrete Approach

In this section, we assume that Assumptions 1 and 2 hold, but also Assumption 3.

Let us recall that Assumption 3, that states a uniform coercivity result for the discrete Gramians Λ_{Th}, is not a consequence of classical convergence results for numerical methods. It rather consists in a very precise result on the dynamics of the discrete equations (1.11) which is the key of the discrete approach.

1.5.1 Proof of Theorem 1.4

Proof (Theorem 1.4). It closely follows the proof of Theorem 1.2, except that now, following the proof of Eq. (1.47), based on Assumption 3, we can prove that, for $\rho \in (0, \rho_2)$ and

$$\delta_d(\rho) = \sqrt{1 - 2\frac{\rho}{\mathscr{C}^2} + \rho^2 \mathscr{C}_{ad}^4}, \tag{1.63}$$

we have that

$$\|(I - \rho \Lambda_{Th})\varphi_h\|_h \leq \delta_d \|\varphi_h\|_h. \tag{1.64}$$

instead of Eq. (1.55).

Consequently, estimate (1.60) can be replaced by

$$\left\|\varphi_{0h}^{k+1} - R_h \varphi_0^{k+1}\right\|_h \leq \delta_d \left\|\varphi_{0h}^k - R_h \varphi_0^k\right\|_h$$
$$+ \rho \|y_{0h} - R_h y_0\|_h + C\rho h^\theta \|y_0\|_s, \quad k \in \mathbb{N}, \tag{1.65}$$

where C is a constant independent of k and $h > 0$.

Of course, this can be rewritten as

$$\frac{1}{\delta_d^{k+1}} \left\|\varphi_{0h}^{k+1} - R_h \varphi_0^{k+1}\right\|_h - \frac{1}{\delta_d^k} \left\|\varphi_{0h}^k - R_h \varphi_0^k\right\|_h$$
$$\leq \frac{1}{\delta_d^{k+1}} \left(\rho \|y_{0h} - R_h y_0\|_h + C\rho h^\theta \|y_0\|_s\right), \tag{1.66}$$

so

$$\frac{1}{\delta_d^k} \left\|\varphi_{0h}^k - R_h \varphi_0^k\right\|_h \leq \left(\sum_{j=1}^k \frac{1}{\delta_d^j}\right) \left(\rho \|y_{0h} - R_h y_0\|_h + C\rho h^\theta \|y_0\|_s\right).$$

Of course, since $\delta_d \in (0, 1)$, by construction, this implies that for all $k \in \mathbb{N}$,

$$\left\|\varphi_{0h}^k - R_h \varphi_0^k\right\|_h \leq \frac{1}{1 - \delta_d} \left(\rho \|y_{0h} - R_h y_0\|_h + C\rho h^\theta \|y_0\|_s\right).$$

Using then Eq. (1.61), estimate (1.39) immediately follows, similarly as in the proof of Theorem 1.2. □

1.5.2 Proof of Theorem 1.5

Proof (Theorem 1.5). Using Eq. (1.39), one only needs to find k such that

$$\left\| \varphi_0^k - \Phi_0 \right\|_X \leq C h^\theta \left\| y_0 \right\|_s.$$

Thus, we only have to check that this estimate holds for any $k \geq K_h^d$, K_h^d given by Eq. (1.40). But this is an immediate consequence of Theorem 1.1. This concludes the proof of Theorem 1.5. $\square$

1.6 Advantages of the Discrete Approach

When comparing the results in Theorems 1.3 and 1.5, one may think that the continuous approach, which applies with a lot of generality, yields essentially the same convergence estimates as the discrete one, more intricate, making the latter irrelevant. This is not the case, and we list below an important number of facts that may be used to compare the two techniques.

1.6.1 The Number of Iterations

A first look on the number of iterations K_h^c, K_h^d in Eqs. (1.35) and (1.40) indicates that they do not depend significantly but only in a logarithmic manner, on the mesh size h. They rather depend essentially on δ given by Eq. (1.20), which is close to 1.

To be more precise, formula (1.35) requires to have an estimate on $\delta(\rho)$, which depends on the observability and admissibility constants in an intricate way, see Eq. (1.20). However, these two constants are not easy to compute in general situations and, usually, one can only get some bounds on them.

Assume that C_{obs} is bounded by $C_{\text{obs,est}}$ and C_{ad} by $C_{\text{ad,est}}$ (here and below, the index "est" stands for estimated). Then, taking $\rho < 2/(C_{\text{ad,est}}^4 C_{\text{obs,est}}^2)$, Theorem 1.3 applies, and $\delta(\rho) \leq \delta_{\text{est}}$, where δ_{est} is defined by

$$\delta_{\text{est}} = \sqrt{1 - \frac{2\rho}{C_{\text{obs,est}}^2} + \rho^2 C_{\text{ad,est}}^4},$$

and therefore

$$\frac{1}{|\log(\delta)|} \leq \frac{1}{|\log(\delta_{\text{est}})|},$$

which means that K_h^c in Eq. (1.35) can only be estimated from above

$$K_h^c \leq K_{\text{est},h}^c := \left\lfloor \theta \frac{\log(h)}{\log(\delta_{\text{est}})} \right\rfloor.$$

Of course, this $K_{\text{est},h}^c$ can be much larger than K_h^c, but according to Eq. (1.62), estimate Eq. (1.36) also holds with that stopping time $K_{\text{est},h}^c$ instead of K_h^c.

Similarly, when applying Theorem 1.5, that is when Assumption 3 holds, one can bound K_h^d in Eq. (1.40) by

$$K_{\text{est},h}^d := \left\lfloor \theta \frac{\log(h)}{\log(\delta_{\text{est}})} - (s+1) \frac{\log(|\log(h)|)}{\log(\delta_{\text{est}})} \right\rfloor.$$

But here, the final iteration time can be any number k larger than $K_{\text{est},h}^d$, and in particular it can be chosen to be $k \simeq \infty$. Hence, in the discrete approach, we do not really care about the estimates we have on K_h^d. This is in contrast with the behavior of the continuous approach in which, taking the number of iterations beyond the optimal threshold, can deteriorate the error estimate and actually makes the method diverge, see Sect. 1.7.

Actually, in the discrete approach, we prove a Γ-convergence result for the minimizers of the functionals J_h in Eq. (1.12) towards that of J in Eq. (1.9). Thus, one can use more sophisticated and rapid algorithms to compute the minimum of J_h as, for instance, conjugate gradient methods; see Sect. 1.6.3. The convergence will then be faster and the number of iterations smaller.

1.6.2 Controlling Non-smooth Data

Here, we are interested in the case in which $y_0 \in X$ and we have some discrete initial data $y_{0h} \in V_h$ such that $E_h y_{0h}$ converge to y_0 strongly in X. Then, neither Theorem 1.3 nor Theorem 1.5 applies.

However, in the discrete approach, that is when supposing Assumption 3, similarly as in Theorem 1.1, we have the following:

Theorem 1.6. *Suppose that Assumptions 1, 2, and 3 are satisfied. Let $h > 0$, $y_{0h} \in V_h$ and Φ_{0h} be the solution of Eq. (1.42).*

For any $\rho \in (0, 2/\mathscr{C}_{ad}^4 \mathscr{C}^2)$, and $\delta(\rho)$ as in Eq. (1.63), the sequence φ_{0h}^k defined by Eq. (1.32) satisfies.

$$\left\| \varphi_{0h}^k - \Phi_{0h} \right\|_h \leq \delta^k \left\| \Phi_{0h} \right\|_h, \quad k \in \mathbb{N}. \tag{1.67}$$

Of course, the proof of Theorem 1.6 closely follows the one of Theorem 1.1 and is therefore omitted.

Note that, since Φ_{0h} is the solution of Eq. (1.42), it coincides with the unique (because of Assumption 3) minimizer of J_h defined in Eq. (1.12), and the iterates φ_{0h}^k defined by Eq. (1.32) simply are those of the steepest descent algorithm for J_h.

But, using Theorem 1.6, we can prove that, if $y_0 \in X$ and $E_h y_{0h}$ converge to y_0 in X, the sequence of $E_h \Phi_{0h}$ converges in X to Φ_0:

Theorem 1.7. *Suppose that Assumptions 1, 2, and 3 are satisfied. Let $y_0 \in X$ and $\Phi_0 \in X$ be the solution of Eq. (1.14). Let $y_{0h} \in V_h$ and $\Phi_{0h} \in V_h$ be the solution of Eq. (1.42).*

If $E_h y_{0h}$ weakly (respectively, strongly) converges to y_0 in X as $h \to 0$, $E_h \Phi_{0h}$ weakly (respectively, strongly) converges to Φ_0 in X.

Theorem 1.7 is actually well known and is usually deduced by suitable convergence results, similarly as in [16].

Proof. Since $E_h y_{0h}$ weakly converges to y_0 in X, it is bounded in X. Therefore, using Eq. (1.43), $E_h \Phi_{0h}$ is bounded in X. Hence it weakly converges to some $\tilde{\phi}_0$ in X.

Using that Φ_0 and Φ_{0h} solve, respectively Eqs. (1.14) and (1.42), for all ψ_0 and ψ_{0h}, we have

$$\langle \Lambda_T \Phi_0, \psi_0 \rangle_X + \langle \psi_0, y_0 \rangle_X = 0, \quad \langle \Lambda_{Th} \Phi_{0h}, \psi_{0h} \rangle_h + \langle \psi_{0h}, y_{0h} \rangle_h = 0. \qquad (1.68)$$

In particular, using that Λ_T and Λ_{Th} are self-adjoint in V and V_h, respectively,

$$\langle \Phi_0, \Lambda_T \psi_0 \rangle_X + \langle \psi_0, y_0 \rangle_X = 0, \quad \langle \Phi_{0h}, \Lambda_{Th} \psi_{0h} \rangle_h + \langle \psi_{0h}, y_{0h} \rangle_h = 0. \qquad (1.69)$$

Let us then fix $\psi_0 \in X_s$ and $\psi_{0h} = R_h \psi_0$. According to Assumption 1,

$$E_h \psi_{0h} \xrightarrow[h \to 0]{} \psi_0 \quad \text{in } X, \quad E_h \Lambda_{Th} \psi_{0h} \xrightarrow[h \to 0]{} \Lambda_T \psi_0 \quad \text{in } X.$$

In particular,

$$\begin{aligned}
\langle \tilde{\phi}_0, \Lambda_T \psi_0 \rangle_X &= \lim_{h \to 0} \langle \Phi_{0h}, \Lambda_{Th} \psi_{0h} \rangle_h = -\lim \langle \psi_{0h}, y_{0h} \rangle_h \\
&= -\langle \psi_0, y_0 \rangle_X = \langle \Phi_0, \Lambda_T \psi_0 \rangle_X.
\end{aligned}$$

Using that Λ_T is an isomorphism on X_s and the fact that X_s is dense in X, we thus deduce that $\tilde{\phi}_0 = \Phi_0$, i.e., $E_h \Phi_{0h}$ weakly converges to Φ_0 in X.

Let us now assume that $E_h y_{0h}$ strongly converges to y_0 in X. Set $\Phi_{0h} = \Lambda_{Th}^{-1} y_{0h}$, $\Phi_0 = \Lambda_T^{-1} y_0$. Let $\varepsilon > 0$. Set $\tilde{y}_0 \in \mathscr{D}(A^s)$ such that $\|y_0 - \tilde{y}_0\|_X \le \varepsilon$. The observability of the continuous model then implies that, setting $\tilde{\Phi}_0 = \Lambda_T^{-1} \tilde{y}_0$, $\|\tilde{\Phi}_0 - \Phi_0\|_X \le C\varepsilon$. Besides, applying Theorem 1.5 to $\tilde{y}_0$, there exists a sequence $\tilde{y}_{0h}$ such that

$$\|E_h \tilde{y}_{0h} - \tilde{y}_0\|_X \le Ch^\theta, \quad \|E_h \tilde{\Phi}_{0h} - \tilde{\Phi}_0\|_X \le Ch^\theta,$$

where $\tilde{\Phi}_{0h} = \Lambda_{Th}^{-1} \tilde{y}_{0h}$.

Finally, since Λ_{Th}^{-1} is uniformly bounded by Assumption 3,

$$\|E_h \tilde{\Phi}_{0h} - E_h \Phi_{0h}\|_X \le C \|E_h \tilde{y}_{0h} - E_h y_{0h}\|_X.$$

But

$$\|E_h \tilde{y}_{0h} - E_h y_{0h}\|_X \leq \|E_h \tilde{y}_{0h} - \tilde{y}_0\|_X + \|\tilde{y}_0 - y_0\|_X + \|y_0 - E_h y_{0h}\|_X$$
$$\leq Ch^\theta + \varepsilon + \|y_0 - E_h y_{0h}\|_X,$$

and thus

$$\|\Phi_0 - E_h \Phi_{0h}\|_X \leq \|\Phi_0 - \tilde{\Phi}_0\|_X + \|\tilde{\Phi}_0 - E_h \tilde{\Phi}_{0h}\|_X + \|E_h \tilde{\Phi}_{0h} - E_h \Phi_{0h}\|_X$$
$$\leq Ch^\theta + C\varepsilon + C\|y_0 - E_h y_{0h}\|_X.$$

This last estimate proves that for all $\varepsilon > 0$,

$$\limsup_{h \to 0} \|\Phi_0 - E_h \Phi_{0h}\|_X \leq C\varepsilon.$$

This concludes the proof of the strong convergence of $E_h \Phi_{0h}$ to Φ_0 as $h \to 0$. □

Of course, one can go even further and analyze if Assumption 3 is really needed to get convergences of the discrete minima Φ_{0h} of J_h towards the continuous one Φ_0 of J. It turns out that Assumption 3 is indeed needed as numerical evidences show; see Sect. 1.7 and [20, 21, 23] and [16, Theorem 8] for a theoretical proof.

To sum up, the discrete approach ensures the convergence of discrete controls even when the initial data to be controlled are only in X, whereas the continuous approach does not work under these low regularity assumptions.

1.6.3 Other Minimization Algorithms

So far we have chosen to use the steepest descent algorithm for the minimization of J in Eq. (1.9). Of course, many other choices yield better convergence results, in particular the conjugate gradient algorithm, when dealing with quadratic coercive functionals.

However, if one uses the conjugate gradient algorithm to minimize the functional J in Eq. (1.9), we do not know if, similarly as in Theorem 1.1, the iterations converge in X_s when the initial data to be controlled are in X_s. To our knowledge, this is an open problem. This is related to the fact that the conjugate gradient algorithm strongly uses orthogonality properties in the natural space X endowed with its natural scalar product $\langle \cdot, \cdot \rangle_X$ and with the scalar product adapted to the minimization problem $\langle \Lambda_T \cdot, \cdot \rangle_X$.

This prevents us from using the conjugate gradient algorithm when following the continuous approach.

However, when considering the discrete approach, since we proved (Theorem 1.5) that the minimizers Φ_{0h} of J_h in Eq. (1.12) converge to the minimizer Φ_0 of J when $E_h y_{0h}$ converge in X, there is full flexibility in the choice of the algorithm to effectively compute the minimizer of J_h. In particular, we can then use the conjugate gradient algorithm, for which we know that the minimum of J_h is attained in at most $\dim(V_h)$ iterations and in general much faster than that.

As shown in the applications below, this makes the discrete approach more efficient for numerics.

1.7 Application to the Wave Equation

Below, we focus on the emblematic example of the wave equation controlled from the boundary or from an open subset.

In particular, we will focus on the case of the $1-d$ wave equation controlled from the boundary, in which case we can easily illustrate our results with some numerical simulations since the control function will simply be a function of time.

We then explain how our approach works in the context of distributed controls so to compare it briefly with the results in [9].

1.7.1 Boundary Control

1.7.1.1 The Continuous Case

Let us consider the $1-d$ wave equation controlled from $x=1$:

$$\begin{cases} \partial_{tt}y - \partial_{xx}y = 0, & (t,x) \in \mathbb{R}_+ \times (0,1), \\ y(t,0) = 0, \; y(t,1) = v(t) & (t,x) \in \mathbb{R}_+, \\ (y(0,x),\partial_t y(0,x)) = (y_0(x),y_1(x)), \; x \in (0,1). \end{cases} \qquad (1.70)$$

Then, set $X = L^2(0,1) \times H^{-1}(0,1)$, A the operator defined by

$$A = \begin{pmatrix} 0 & I \\ \partial_{xx}^D & 0 \end{pmatrix}, \quad \mathscr{D}(A) = H_0^1(0,1) \times L^2(0,1),$$

where ∂_{xx}^D is the Laplace operator defined on $H^{-1}(0,1)$ with domain $\mathscr{D}(\partial_{xx}^D) = H_0^1(0,1)$ (in other words, ∂_{xx}^D is the Laplacian with Dirichlet boundary conditions) and B the operator defined by

$$Bv = \begin{pmatrix} 0 \\ -\partial_{xx}^D \tilde{y} \end{pmatrix}, \quad \text{with } \tilde{y} \text{ solving } \begin{cases} -\partial_{xx}\tilde{y} = 0, & x \in (0,1), \\ \tilde{y}(0) = 0, \; \tilde{y}(1) = v. \end{cases}$$

Here, endowing $L^2(0,1)$ with its usual L^2-norm and the space $H^{-1}(0,1)$ with the norm $\|(-\partial_{xx}^D)^{-1/2} \cdot \|_{L^2}$, A is skew-adjoint on $L^2(0,1) \times H^{-1}(0,1)$ and B is an admissible control operator. We refer to [49, Sect. 9.3] (see also [33, 35]) for the proof of these facts.

We can then consider the adjoint equation

$$
\begin{cases}
\partial_{tt}\varphi - \partial_{xx}\varphi = 0, & (t,x) \in \mathbb{R}_+ \times (0,1), \\
\varphi(t,0) = 0 = \varphi(t,1), & (t,x) \in \mathbb{R}_+, \\
(\varphi(0,x), \partial_t \varphi(0,x)) = (\varphi_0(x), \varphi_1(x)), & x \in (0,1),
\end{cases}
\tag{1.71}
$$

with $(\varphi^0, \varphi^1) \in L^2(0,1) \times H^{-1}(0,1)$. The corresponding admissibility and observability properties (1.3) and (1.5) we need read as follows (see [49, Proposition 9.3.3] for the computation of B^*):

$$
\frac{1}{C}\|(\varphi_0,\varphi_1)\|^2_{L^2 \times H^{-1}} \le \int_0^T |\partial_x[(-\partial^D_{xx})^{-1}\partial_t\varphi](t,1)|^2\, dt \le C\|(\varphi_0,\varphi_1)\|^2_{L^2 \times H^{-1}}.
$$

Of course, when considering these estimates, one easily understands that rather than considering trajectories φ of Eq. (1.71) for initial data $(\varphi_0, \varphi_1) \in L^2(0,1) \times H^{-1}(0,1)$, it is easier to directly work on the set of trajectories $(-\partial^D_{xx})^{-1}\partial_t\varphi$. But this set coincides with the set of trajectories φ of Eq. (1.71) with initial conditions $(\varphi^0, \varphi^1) \in H^1_0(0,1) \times L^2(0,1)$.

Therefore, in the following, we shall only consider solutions φ of Eq. (1.71) with initial data in $H^1_0(0,1) \times L^2(0,1)$.

Also note that this space is the natural one when identifying $L^2(0,1)$ with its dual since the control system (1.70) takes place in $X = L^2(0,1) \times H^{-1}(0,1)$, and therefore $X^* = H^1_0(0,1) \times L^2(0,1)$. This is the usual duality setting in Lions [36], but the above argument ensures that all the results of this article, which have been obtained within the setting of abstract conservative systems on Hilbert spaces identified with their duals, can also be applied in the context of the usual duality pairing between $X = L^2(0,1) \times H^{-1}(0,1)$ and $X^* = H^1_0(0,1) \times L^2(0,1)$.

Also note that the duality then reads as follows:

$$
\langle (y_0,y_1),(\varphi_0,\varphi_1) \rangle_{(L^2 \times H^{-1}),(H^1_0 \times L^2)} = -\int_0^1 y_0\varphi_1 + \int_0^1 \partial_x(-\partial^D_{xx})^{-1}y_1\partial_x\varphi_0.
$$

In that context, the relevant counterparts of Eqs. (1.3) and (1.5) are then given by

$$
\frac{1}{C}\|(\varphi_0,\varphi_1)\|^2_{H^1_0 \times L^2} \le \int_0^T |\partial_x\varphi(t,1)|^2\, dt \le C\|(\varphi_0,\varphi_1)\|^2_{H^1_0 \times L^2}.
\tag{1.72}
$$

Such a result is well known to hold if and only if $T \ge 2$; see [36]. This can actually be proved very easily solving the wave equation (1.71) using Fourier series and Parseval's identity.

Therefore, in the sequel, we take $T > 2$ and η as in Eq. (1.6) with $T^* = 2$. To be more precise, η will be chosen such that η is $C^1([0,T])$ and satisfies $\eta(0) = \eta(T) = \eta'(0) = \eta'(T) = 0$.

The corresponding functional Eq. (1.9) is then defined on $H^1_0(0,1) \times L^2(0,1)$ as follows:

$$
J(\varphi_0,\varphi_1) = \frac{1}{2}\int_0^T \eta(t)|\partial_x\varphi(t,1)|^2\, dt + \langle (y_0,y_1),(\varphi_0,\varphi_1) \rangle_{(L^2 \times H^{-1}),(H^1_0 \times L^2)}.
\tag{1.73}
$$

The corresponding Gramian operator Λ_T is then given as follows: For $(\varphi_0, \varphi_1) \in H_0^1(0,1) \times L^2(0,1)$, solve

$$
\begin{cases}
\partial_{tt}\varphi - \partial_{xx}\varphi = 0, & (t,x) \in (0,T) \times (0,1), \\
\varphi(t,0) = \varphi(t,1) = 0, & t \in (0,T), \\
(\varphi(0,\cdot), \partial_t\varphi(0,\cdot)) = (\varphi_0, \varphi_1), & x \in (0,1).
\end{cases}
\tag{1.74}
$$

Then solve

$$
\begin{cases}
\partial_{tt}\psi - \partial_{xx}\psi = 0, & (t,x) \in (0,T) \times (0,1), \\
\psi(t,0) = 0, \ \psi(t,1) = -\eta(t)\partial_x\varphi(t,1), & t \in (0,T), \\
(\psi(T,\cdot), \partial_t\psi(T,\cdot)) = (0,0), & x \in (0,1).
\end{cases}
\tag{1.75}
$$

Then

$$
\Lambda_T(\varphi_0, \varphi_1) = ((-\partial_{xx}^D)^{-1}\partial_t\psi(0,\cdot), -\psi(0,\cdot)).
\tag{1.76}
$$

Note that the solution ψ of Eq. (1.75) is a solution by transposition and belongs to the space $C^0([0,T];L^2(0,1)) \cap C^1([0,T];H^{-1}(0,1))$ since its boundary data only belongs to $L^2(0,T)$. Therefore, when computing Λ_T, we have to identify $L^2(0,1) \times H^{-1}(0,1)$ as the dual of $H_0^1(0,1) \times L^2(0,1)$ as explained in the previous paragraph, i. e., using the map

$$
L^2(0,1) \times H^{-1}(0,1) \to H_0^1(0,1) \times L^2(0,1) : (\psi_0, \psi_1) \mapsto ((-\partial_{xx}^D)^{-1}\psi_1, -\psi_0).
$$

The Continuous Setting

We are then in position to write the algorithm of Theorem 1.1 in the continuous setting:

Step 0: Set $(\varphi_0^0, \varphi_1^0) = (0,0)$.

The induction formula $k \to k+1$: For $k \geq 0$, set $(\varphi_0^{k+1}, \varphi_1^{k+1})$ as

$$
(\varphi_0^{k+1}, \varphi_1^{k+1}) = (I - \rho\Lambda_T)(\varphi_0^k, \varphi_1^k) - \rho((-\partial_{xx}^D)^{-1}y_1, -y_0).
\tag{1.77}
$$

Note that the control function is then approximated by the sequence

$$
v^k(t) = \eta(t)\partial_x\varphi^k(t,1), \quad t \in (0,T),
\tag{1.78}
$$

where φ^k is the solution of Eq. (1.74) with initial data $(\varphi_0^k, \varphi_1^k)$. Indeed, formula (1.10) then reads as follows: if (Φ_0, Φ_1) denotes the minimum of J in Eq. (1.73), then the control function V that controls Eq. (1.70) and that minimizes the $L^2(0,T;dt/\eta)$-norm among all admissible controls is given by

$$V(t) = \eta(t)\partial_x \Phi(t,1), \quad t \in (0,T), \tag{1.79}$$

where Φ is the solution of Eq. (1.74) with initial data (Φ_0, Φ_1).

1.7.1.2 The Continuous Approach

Theoretical Setting

Here, we discretize the wave equations (1.74) and (1.75) using the finite-difference approximation of the Laplace operator on a uniform mesh of size $h > 0$, $h = 1/(N+1)$ with $N \in \mathbb{N}$. Below, $\varphi_{j,h}, \psi_{j,h}$ are, respectively, the approximations of φ, ψ solutions of Eq. (1.74)–(1.75) at the point jh. We shall also make use of the notation φ_h, ψ_h to denote, respectively, the N-component vectors with coordinates $\varphi_{j,h}, \psi_{j,h}$.

We shall thus introduce the following discrete version of the Gramian operator. Given $(\varphi_{0h}, \varphi_{1h})$, compute the solution φ_h of the following system:

$$\begin{cases} \partial_{tt}\varphi_{j,h} - \dfrac{1}{h^2}\left(\varphi_{j+1,h} - 2\varphi_{j,h} + \varphi_{j-1,h}\right) = 0, & (t,j) \in (0,T) \times \{1,\cdots,N\}, \\ \varphi_{0,h}(t) = \varphi_{N+1,h}(t) = 0, & t \in (0,T), \\ (\varphi_h(0), \partial_t \varphi_h(0)) = (\varphi_{0h}, \varphi_{1h}). \end{cases} \tag{1.80}$$

Then compute the solution ψ_h of the following approximation of Eq. (1.75):

$$\begin{cases} \partial_{tt}\psi_{j,h} - \dfrac{1}{h^2}\left(\psi_{j+1,h} - 2\psi_{j,h} + \psi_{j-1,h}\right) = 0, & (t,j) \in (0,T) \times \{1,\cdots,N\}, \\ \psi_{0,h}(t) = 0, \; \psi_{N+1,h}(t) = \eta(t)\dfrac{\varphi_{N,h}}{h}, & t \in (0,T), \\ (\psi_h(T), \partial_t \psi_h(T)) = (0,0). \end{cases}$$
$$\tag{1.81}$$

Finally, set Λ_{Th} as

$$\Lambda_{Th}(\varphi_{0h}, \varphi_{1h}) = ((-\Delta_h)^{-1}\partial_t \psi_h(0), -\psi(0)), \tag{1.82}$$

where $u_h = (-\Delta_h)^{-1}f_h$ is the unique solution of the discrete elliptic problem

$$\begin{cases} -\dfrac{1}{h^2}\left(u_{j+1,h} - 2u_{j,h} + u_{j-1,h}\right) = f_{j,h}, \; j \in \{1,\cdots,N\}, \\ u_{0,h} = u_{N+1,h} = 0. \end{cases}$$

The continuous approach then reads as follows:

Step 0: Set $(\varphi_{0h}^{0,c}, \varphi_{1h}^{0,c}) = (0,0)$.

The induction formula $k \to k+1$: For $k \geq 0$, set $(\varphi_{0h}^{k+1,c}, \varphi_{1h}^{k+1,c})$ as

$$(\varphi_{0h}^{k+1,c}, \varphi_{1h}^{k+1,c}) = (I - \rho\Lambda_{Th})(\varphi_{0h}^{k,c}, \varphi_{1h}^{k,c}) - \rho((-\Delta_h)^{-1}y_{1h}, -y_{0h}). \tag{1.83}$$

The superscript c is here to emphasize that this is the sequence computed by the continuous approach.

Let us check that this scheme fits the abstract setting of Theorem 1.3. In particular, to use Theorem 1.3, V_h, E_h, and R_h need to be defined and Assumptions 1 and 2 verified:

- $V_h = \mathbb{R}^N \times \mathbb{R}^N$, where the first N components correspond to the approximation of the displacement and the last N ones to the velocity.
- To a discrete vector $\varphi_h \in \mathbb{R}^N$, there exists a unique family of Fourier coefficients $(\hat{a}_k[\varphi_h])_{k \in \mathbb{N}}$ such that

$$\varphi_{j,h} = \sqrt{2} \sum_{k=1}^{N} \hat{a}_k[\varphi_h] \sin(k\pi jh), \quad j \in \{1, \cdots, N\}.$$

This is due to the fact that the family of vectors

$$w_h^k = (\sqrt{2} \sin(k\pi jh))_{j \in \{1, \cdots, N\}}$$

forms a basis of $\mathbb{R}^N$ endowed with the scalar product $h\langle \cdot, \cdot \rangle_{\mathbb{R}^N}$; see Chap. 2, Sect. 2.2.

Then we introduce the following continuous extension:

$$e_h \varphi_h(x) = \sqrt{2} \sum_{k=1}^{N} \hat{a}_k[\varphi_h] \sin(k\pi x), \quad x \in (0,1)$$

and set $E_h = \text{diag}(e_h, e_h)$. This extension will be extensively studied in Sect. 3.2, where it is denoted by $\mathbb{F}_h$.

The corresponding norm on V_h is given by

$$\left\| \begin{pmatrix} \varphi_{0h} \\ \varphi_{1h} \end{pmatrix} \right\|_h^2 = \sum_{k=1}^{N} \left(k^2 \pi^2 |\hat{a}_k[\varphi_{0h}]|^2 + |\hat{a}_k[\varphi_{1h}]|^2 \right), \tag{1.84}$$

which is equivalent (see Chap. 3, Sect. 3.2) to the classical discrete energy of Eq. (1.80), given by

$$E_h[\varphi_h] = h \sum_{j=0}^{N} \left(\frac{\varphi_{j+1,0h} - \varphi_{j,0h}}{h} \right)^2 + h \sum_{j=1}^{N} |\varphi_{j,1h}|^2. \tag{1.85}$$

The operator A_h defined by

$$A_h = \begin{pmatrix} 0 & Id_{\mathbb{R}^N} \\ \Delta_h & 0 \end{pmatrix},$$

where Δ_h denotes the $N \times N$ matrix taking value $-2/h^2$ on the diagonal and $1/h^2$ on the upper and lower diagonals, is skew-adjoint with respect to the scalar product of V_h. Of course, this operator A_h is the one corresponding to system (1.80) in the sense that φ_h solves Eq. (1.80) if and only if

$$\partial_t \begin{pmatrix} \varphi_h \\ \partial_t \varphi_h \end{pmatrix} = A_h \begin{pmatrix} \varphi_h \\ \partial_t \varphi_h \end{pmatrix}, \quad t \in (0,T).$$

Also note that the fact that A_h is skew-adjoint implies that solutions φ_h of Eq. (1.80) have constant (with respect to time) V_h norms. This quantity is usually called the discrete energy of the solutions of Eq. (1.80).

The operator $B_h B_h^*$ is now simply given by

$$B_h B_h^* \begin{pmatrix} \varphi_{0h} \\ \varphi_{1h} \end{pmatrix} = \begin{pmatrix} 0 \\ f_h \end{pmatrix},$$

where $f_h \in \mathbb{R}^N$ is such that its $N-1$ first components vanish and whose N-th component is $-\varphi_{N,0h}/h^3$.

The operator R_h on $X = H_0^1(0,1) \times L^2(0,1)$ has a diagonal form $\text{diag}(r_h, r_h)$, where $r_h : H^{-1}(0,1) \longrightarrow \mathbb{R}^N$ is defined as follows: for $\varphi \in H^{-1}(0,1)$, expand it into its Fourier series

$$\varphi(x) = \sqrt{2} \sum_{k=1}^{\infty} \hat{a}_k \sin(k\pi x), \quad x \in (0,1),$$

and then set $r_h \varphi \in \mathbb{R}^N$ as

$$(r_h \varphi)_j = \sqrt{2} \sum_{k=1}^{N} \hat{a}_k \sin(k\pi jh), \quad j \in \{1, \cdots, N\}.$$

Let us now check Assumptions 1 and 2.

Assumption 1. Estimates (1.26)–(1.27) are very classical with $s = 1$ and $\theta = 1$ (see, e.g., [4]): There exists a constant C such that for all $(\varphi_0, \varphi_1) \in H^2 \cap H_0^1(0,1) \times H_0^1(0,1)$,

$$\|(E_h R_h - Id)(\varphi_0, \varphi_1)\|_{H_0^1 \times L^2} \leq Ch \|(\varphi_0, \varphi_1)\|_{H^2 \cap H_0^1 \times H_0^1}. \tag{1.86}$$

As already mentioned, estimate (1.28) is a consequence of Assumption 2 with Eq. (1.26).

To show Eq. (1.29) we take the initial data $(\varphi_0, \varphi_1) \in H^2 \cap H_0^1(0,1) \times H^1(0,1)$ and denote by φ the corresponding solution of Eq. (1.74). Then, taking $(\varphi_{0h}, \varphi_{1h}) = R_h(\varphi_0, \varphi_1)$ and φ_h the corresponding solution of Eq. (1.80), from Proposition 3.7 in Chap. 3, we get

$$\left\| \partial_x \varphi(t,1) + \frac{\varphi_{N,h}}{h} \right\|_{L^2(0,T)} \leq Ch^{2/3} \|(\varphi_0, \varphi_1)\|_{H^2 \cap H_0^1 \times H_0^1}. \tag{1.87}$$

Thus, according to the convergence results of the numerical scheme (1.81) to Eq. (1.75) in Proposition 4.8 in Chap. 4, setting ψ_h the solution of Eq. (1.81) with boundary data $\eta \varphi_{N,h}/h$ and ψ the solution of Eq. (1.75) with boundary data $-\eta \partial_x \varphi(1,t)$, we obtain

$$\|(e_h(\psi_h(0)), e_h(\partial_t \psi_h(0))) - (\psi(0), \partial_t \psi(0))\|_{L^2 \times H^{-1}} \leq Ch^{2/3} \|(\varphi_0, \varphi_1)\|_{H^2 \cap H_0^1 \times H_0^1}.$$

Then, since $(\psi(0), \partial_t \psi(0)) \in H_0^1(0,1) \times L^2(0,1)$ because of the fact that $(\varphi_0, \varphi_1) \in H^2 \cap H_0^1(0,1) \times H_0^1(0,1)$, see Sect. 4.2 in Chap. 4,

$$\left\| (e_h(\partial_t \psi_h(0)), e_h((-\Delta_h)^{-1} \psi_h(0))) - (\partial_t \psi(0), (-\partial_{xx}^D)^{-1} \partial_t \psi(0)) \right\|_{H_0^1 \times L^2}$$
$$\leq Ch^{2/3} \left\| (\varphi_0, \varphi_1) \right\|_{H^2 \cap H_0^1 \times H_0^1},$$

which proves

$$\left\| (E_h \Lambda_{Th} R_h - \Lambda_T)(\varphi_0, \varphi_1) \right\|_{H_0^1 \times L^2} \leq Ch^{2/3} \left\| (\varphi_0, \varphi_1) \right\|_{H^2 \cap H_0^1 \times H_0^1},$$

and then Eq. (1.29) with $\theta = 2/3$ and $s = 1$ since $\Lambda_T(\varphi_0, \varphi_1) \in H^2 \cap H_0^1 \times H_0^1$ and then Eq. (1.86) applies.

Assumption 2. The uniform admissibility result is ensured by the fact that the map $(\varphi_{0h}, \varphi_{1h}) \in V_h \mapsto -\varphi_{N,h}/h \in L^2(0,T)$, where φ_h is the solution of Eq. (1.80), is bounded, uniformly with respect to $h > 0$. This is a simple consequence of the multiplier identity given in Lemma 2.2 in [28], see also Chap. 2, Theorem 2.1.

Besides, for $b_h \in L^2(0,T)$, the solution ψ_h of

$$\begin{cases} \partial_{tt} \psi_{j,h} - \dfrac{1}{h^2}\left(\psi_{j+1,h} - 2\psi_{j,h} + \psi_{j-1,h} \right) = 0, \\ \hspace{4cm} (t,j) \in (0,T) \times \{1, \cdots, N\}, \\ \psi_{0,h}(t) = 0, \ \psi_{N+1,h}(t) = b_h, \hspace{1.5cm} t \in (0,T), \\ (\psi_h(T), \partial_t \psi_h(T)) = (0,0), \end{cases}$$

is such that the $L^2(0,1) \times H^{-1}(0,1)$-norm of $(e_h(\psi_h(0)), e_h(\partial_t \psi_h(0)))$ is bounded, uniformly with respect to $h > 0$, by the $L^2(0,T)$-norm of its boundary term b_h, see Chap. 4, Theorem 4.6.

Finally, one easily checks that, for some constants C independent of $h > 0$, see Chap. 4, Sect. 4.2,

$$\left\| (e_h(-\Delta_h)^{-1}\psi_{1h}, e_h\psi_{0h}) \right\|_{H_0^1 \times L^2} \leq C \left\| (e_h(\psi_{0h}), e_h(\psi_{1h})) \right\|_{L^2 \times H^{-1}}.$$

Assumption 3. Note that, in that setting, Assumption 3 does not hold. Indeed, the numerical scheme under consideration generates spurious high-frequency waves that travel at arbitrarily small velocity (see [16, 21, 28, 48, 52]) that cannot be observed. Therefore, the discrete systems (1.80) are not uniformly observable, whatever $T > 0$ is.

We are thus in a situation in which Theorem 1.3 applies with $s = 1$ and $\theta = 2/3$. We now illustrate these results by some numerical experiments.

Remark 1.2. Using Propositions 3.7 and 4.9 and Theorem 4.4, one can prove that Assumption 1 actually holds for any $s \in (0,3]$ with $\theta = 2s/3$ and thus Theorem 1.3 applies for any $s \in (0,3]$.

Numerical Simulations

To apply our numerical method, we need estimates on C_{obs} and C_{ad}. In this $1-d$ context, it is rather easy to get good approximations, since for any solution φ of Eq. (1.74), using Fourier series and Parseval's identity, we get

$$\int_0^2 |\partial_x \varphi(t,1)|^2 \, dt = 2 \, \|(\varphi_0, \varphi_1)\|_{H_0^1 \times L^2}^2 \,.$$

Therefore, we can take $T^* = 2$, and we choose $T = 4$. We then have the estimates $C_{\text{obs}}^2 = 1/2$ and $C_{\text{ad}}^2 = 4$. With $\rho = 1/8$, we have $\delta(\rho) = \sqrt{3}/2 \simeq 0.86$.

But ρ should also be smaller than $2/\mathscr{C}_{\text{ad}}^2$, where $\mathscr{C}_{\text{ad}}^2$ is the uniform constant of admissibility in Assumption 2. Using the multiplier method on the discrete equations (1.80) (see Chap. 2, Theorem 2.1) we have $\mathscr{C}_{\text{ad}}^2 \leq 6$ (actually $\mathscr{C}_{\text{ad}}^2(T) \leq T+2$). Since $1/8 \leq 2/6$, ρ_1 in Eq. (1.31) is greater than $1/8$ and then $\rho = 1/8$ is admissible.

In order to test our numerical method, we fix the initial data to be controlled as $y_0 = 0$ and y_1 as follows:

$$y_1(x) = \begin{cases} -1 \text{ on } (0, 1/4), \\ 1 \text{ on } (1/4, 1/2), \\ 2(x-1) \text{ on } (1/2, 1), \end{cases} \tag{1.88}$$

so that we obviously have $(y_0, y_1) \in H_0^1(0,1) \times L^2(0,1)$. In Fig. 1.1, we plot the graph of the initial velocity y_1.

In the numerical simulations below, we represent the functions $v_h^{k,c}$ given, for $k \in \mathbb{N}$, by

$$v_h^{k,c}(t) = -\eta \frac{\varphi_{N,h}^{k,c}(t)}{h}, \quad t \in (0,4),$$

where $\varphi_h^{k,c}$ is the solution of Eq. (1.80) with initial data $(\varphi_{0h}^{k,c}, \varphi_{1h}^{k,c})$, the k-th iterate of the algorithm in the continuous approach.

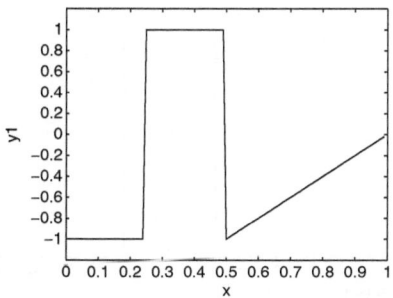

Fig. 1.1 The initial velocity y_1 to be controlled.

For $h = 1/100$, the number of iterations predicted by our method is 21. In Fig. 1.2 (left), we show the control this yields.

To compare the obtained result with the one that the discrete approach yields, we have computed a reference control v_{ref}, using the discrete approach (see Fig. 1.10 for further details) for a much smaller $h = 1/300$. The obtained reference control is plotted in Fig. 1.2 (right).

To better illustrate how the iterative process evolves, we have run it during $50,000$ iterations and drawn the graph of the relative error

$$\left\|v_h^{k,c} - v_{\text{ref}}\right\|_{L^2} / \left\|v_{\text{ref}}\right\|_{L^2}.$$

This is represented in Fig. 1.3. As we see, the error does not reach zero but rather stays bounded from below.

When looking more closely at the evolution of the error, we see that it first decays and then increases.

The smallest error (among the first thousand iterations) is achieved when $k = 13$, which is close to the predicted one. The control obtained for $k = 13$ is plotted in Fig. 1.4, the corresponding error being 6.18%, to be compared with the error at our predicted iteration number ($k = 21$), which is 6.24%.

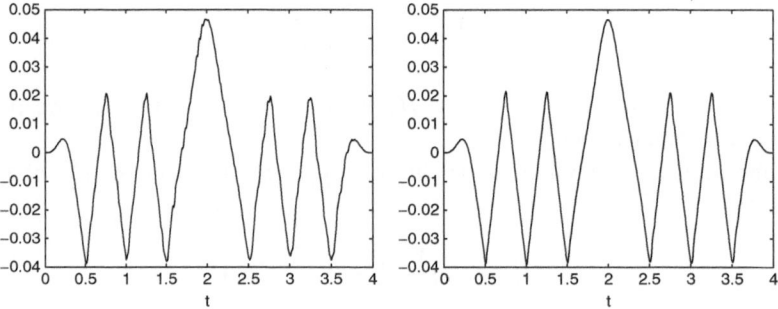

Fig. 1.2 *Left*, the control obtained by the continuous approach at the predicted number of iterations, 21, for $h = 1/100$, $\rho = 1/8$. *Right*, the reference control v_{ref} computed through the discrete approach with $h = 1/300$. The relative error $\left\|v_h^{k=21,c} - v_{\text{ref}}\right\|_{L^2} / \left\|v_{\text{ref}}\right\|_{L^2}$ is 6.24%.

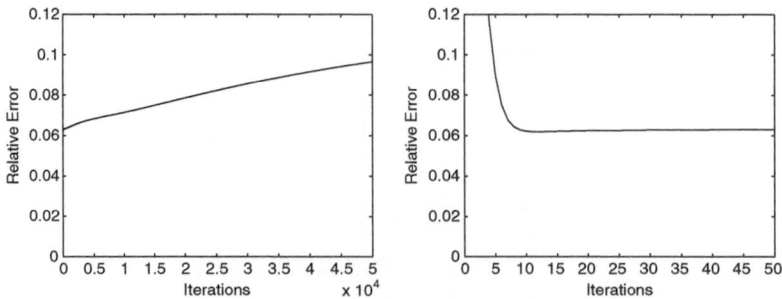

Fig. 1.3 The relative error $\left\|v_h^{k,c} - v_{\text{ref}}\right\|_{L^2} / \left\|v_{\text{ref}}\right\|_{L^2}$ for the continuous approach at each iteration for $h = 1/100$, $\rho = 1/8$. *Left*: iterations from 0 to $50,000$. *Right*: zoom on the iterations between 0 and 50.

The algorithm produces similar results for different values of ρ. For instance, taking $\rho = 0.01$, the predicted iteration number is $k = 156$, and the best iteration turns out to be $k = 180$, yielding a control that looks very much as the one before, the relative error being 6.19%. This confirms, in particular, that the smaller ρ is, the larger is the number of iterations.

It is important to underline that *the limit of the iterative process as the number of iterations tends to infinity, $k \to \infty$, converges to the control of the semi-discrete dynamics, minimizer of the corresponding functional J_h*, defined by

$$J_h(\varphi_{0h}, \varphi_{1h}) = \frac{1}{2} \int_0^T \eta \left| \frac{\varphi_{N,h}}{h} \right|^2 dt - h \sum_{j=1}^N y_{j,0h} \varphi_{j,1h} + h \sum_{j=1}^N y_{j,1h} \varphi_{j,1h},$$

where φ_h is the solution of Eq. (1.80) with initial data $(\varphi_{0h}, \varphi_{1h})$.

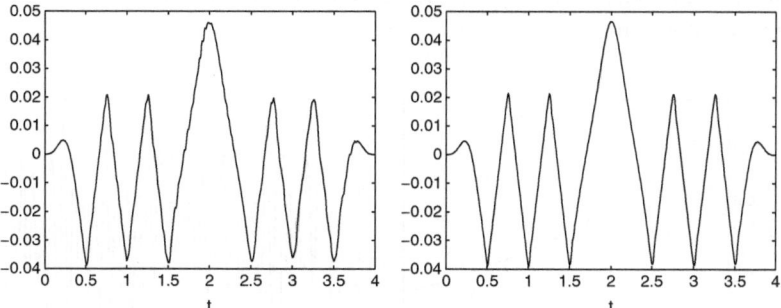

Fig. 1.4 *Left*: the best control obtained by iterating the algorithm of the continuous approach with $h = 1/100$, $\rho = 1/8$ and $k = 13$. *Right*: the reference control. Relative error: $\left\| v_h^{k=13,c} - v_{\text{ref}} \right\|_{L^2} / \|v_{\text{ref}}\|_{L^2} = 6.18\%$.

The exact control of the semi-discrete dynamics is given by the minimizer $(\Phi_{0h}^c, \Phi_{1h}^c)$ of the functional J_h above through the formula

$$v_{*,h}^c = -\eta \frac{\Phi_{N,h}^c}{h}, \quad t \in (0,4),$$

where Φ_h^c is the solution of Eq. (1.80) with initial data $(\Phi_{0h}^c, \Phi_{1h}^c)$.

Note that the Gramians Λ_{Th}, defined by

$$\langle \Lambda_{Th}(\varphi_{0h}, \varphi_{1h}), (\varphi_{0h}, \varphi_{1h}) \rangle_h = \int_0^T \eta \left| \frac{\varphi_{N,h}}{h} \right|^2 dt,$$

are not uniformly coercive with respect to $h > 0$ and their conditioning number degenerates as $\exp(c/h)$ ($c > 0$) as $h \to 0$ (see [39]) and thus the functionals J_h are very ill-conditioned. Therefore, the conjugate gradient algorithm for the minimization of J_h ends up diverging when h is too small.

We take $N = 20$ so that the conjugate gradient algorithm converges. This might seem ridiculously small, but as we said, the conditioning number of the discrete Gramian blows up as $\exp(c/h)$, and numerical experiments show that the conjugate gradient algorithm completely diverges for $N \geq 30$.

For $N = 20$, we can compute the minimizer of the functional J_h using the conjugate gradient algorithm. The corresponding discrete exact control $v^c_{*,h} = \eta \, \Phi^c_{N,h}/h$ is plotted in Fig. 1.5 (right). As one sees, this exact control $v^c_{*,h}$ has a strong spurious oscillating behavior, see for instance the reference control in Fig. 1.4 (right). The relative errors between the iterated controls v^k_h and this limit oscillating control $v^c_{*,h}$ is plotted in Fig. 1.5 (left), exhibiting a slow convergence rate due to the bad conditioning of the Gramian matrix.

These facts constitute a serious *warning about the continuous algorithm*. In particular, if the algorithm is employed for a too large number of iterations k, something that can easily happen since the threshold in the number of iterations may be hard to establish in practice, the corresponding control may be very far away from the actual continuous one.

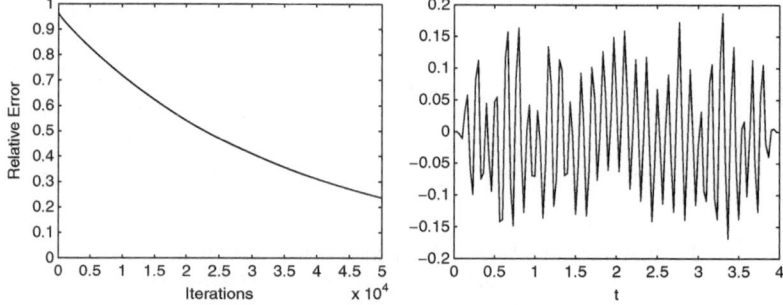

Fig. 1.5 *Left*, the relative error $\left\| v^{k,c}_h - v^c_{*,h} \right\| / \left\| v^c_{*,h} \right\|$ for the continuous approach at each iteration k from $k = 0$ to $k = 50,000$ for $h = 1/20$, $\rho = 1/8$. *Right*, the discrete exact control $v^c_{*,h}$ for $h = 1/20$.

We conclude illustrating the convergence of the continuous algorithm as $h \to 0$. In Fig. 1.6, we plot $\log \left(\left\| v^{K^c_h,c}_h - v_{\text{ref}} \right\| \right)$ versus $|\log(h)|$. By linear regression we get the slope -1.01, which is better than the predicted one, -0.66. This is due to the fact that y_1 is in $H^{-1+s}(0,1)$ for all $s < 3/2$; hence the convergence is expected to be better than $h^{2s/3}$ for all $s < 1/2$ see Remark 1.2.

1.7.1.3 The Discrete Approach

The Theoretical Setting

To build numerical schemes satisfying Assumption 3, one should better understand the dynamics of the solutions of the discrete numerical methods. As observed in [28] and in the numerical tests above, Assumption 3 does not hold when simply

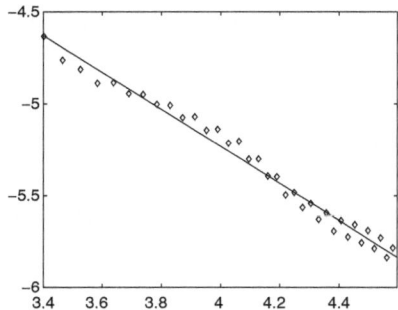

Fig. 1.6 Convergence of the continuous approach as $h \to 0$: $\log\left(\left\|v_h^{K_h^c,c} - v_{\text{ref}}\right\|\right)$ versus $|\log(h)|$, with v_{ref} as in Fig. 1.10 (*right*). The plot is done for $h \in (1/100, 1/30)$, the slope being -1.01.

discretizing the wave equation using a finite difference space semi-discretization. As one sees in Fig. 1.5, the discrete exact control $v_{*,h}^c$ is very far away from the control of the continuous wave equation, for which a good approximation is given by v_{ref}, see Fig. 1.4 (right).

This phenomenon is due to spurious high-frequency numerical waves. To avoid these spurious oscillations one needs to work on filtered subspaces of $V_h = \mathbb{R}^N \times \mathbb{R}^N$.

For instance, for $\gamma \in (0,1)$, consider the filtered space

$$\mathscr{V}_h(\gamma/h) = \left\{ (\varphi_{0h}, \varphi_{1h}), \text{ s.t. } \varphi_{0h}, \varphi_{1h} \in \underset{kh \leq \gamma}{\text{Span}}\left\{ (\sin(k\pi jh))_{j \in \{1,\cdots,N\}} \right\} \right\}.$$

Of course, $\mathscr{V}_h(\gamma/h)$ is a subspace of V_h. Since the functions $(w^k)_k$ (defined by $w_j^k = \sqrt{2}\sin(k\pi jh))$) are eigenfunctions of the discrete Laplace operator (see Sect. 2.2), we can introduce the orthogonal projection P_h^γ of V_h onto $\mathscr{V}_h(\gamma/h)$ (with respect to the scalar product of V_h introduced in Eq. (1.84)) and the Gramian operator

$$\Lambda_{Th}^\gamma = P_h^\gamma \Lambda_{Th} P_h^\gamma. \tag{1.89}$$

The filtering operator P_h^γ simply consists of doing a discrete Fourier transform and then removing the coefficients corresponding to frequency numbers k larger than γ/h.

Assumptions 1 and 2 then hold for any $\gamma \in (0,1)$, with proofs similar to those in the continuous approach. Furthermore, using the results of [28], it can be shown that Assumption 3 also holds when the time T is greater than $T_\gamma := 2/\cos(\pi\gamma/2)$. Note that this is not a consequence of the convergence of the numerical schemes, and this requires a thorough study of the discrete dynamics. The proof of [28] uses a spectral decomposition of the solutions of the discrete wave equation (1.80) and the Ingham inequality for nonharmonic Fourier series. We shall revisit and slightly improve these results in Chap. 2, Theorem 2.1 to get better estimates on the observability constant.

The algorithm that the discrete method yields can then be developed as follows:

Step 0: Set $(\varphi_{0h}^{0,d}, \varphi_{1h}^{0,d}) = (0,0)$.

The induction formula $k \to k+1$: For $k \geq 0$, set $(\varphi_{0h}^{k+1}, \varphi_{1h}^{k+1})$ as

$$(\varphi_{0h}^{k+1,d}, \varphi_{1h}^{k+1,d}) = (I - \rho \Lambda_{Th}^{\gamma})(\varphi_{0h}^{k,d}, \varphi_{1h}^{k,d}) - \rho P_h^{\gamma}((-\Delta_h)^{-1} y_{1h}, -y_{0h}). \quad (1.90)$$

The new algorithm is very similar to the one that the continuous approach yields. The only essential difference is that, now, we have introduced a filtered Gramian matrix Λ_{Th}^{γ} instead of the operator Λ_{Th} used in the continuous approach, in which no filtering appears. However, as we shall see below, this new algorithm is much better behaved.

From now on, we set the filtering parameter $\gamma = 1/3$ and $T = 4$, which is larger than the minimal required time $2/\cos(\pi\gamma/2) = 2/\cos(\pi/6) = 4/\sqrt{3}$ to control the semi-discrete dynamics. The controls that the discrete iterative algorithm yields are

$$v_h^{k,d}(t) = -\eta(t) \frac{\varphi_{N,h}^{k,d}(t)}{h}, \quad t \in (0,4). \quad (1.91)$$

Numerical Simulations

We first need an estimate on the constant of uniform observability. The most explicit one we are aware of is the one given by the multiplier method, given hereafter in Chap. 2, Theorem 2.1, which yields

$$\mathscr{C}_{\mathrm{obs},T^*}^2 = \left(T \cos^2\left(\frac{\gamma\pi}{2}\right) - 2\cos\left(\frac{\pi\gamma}{2}\right) - \frac{h_0}{2} \right)^{-1}, \quad (1.92)$$

where h_0 is the largest mesh size under consideration, and thus, since the function $\eta(t)$ equals to 1 on an interval of length close to 4, one can take the approximation:

$$\mathscr{C}^2(T = 4) \simeq \frac{1}{\sqrt{3}(\sqrt{3} - 1)}.$$

Of course, $\mathscr{C}_{\mathrm{ad}}$ can still be approximated as before by $\mathscr{C}_{\mathrm{ad}}^2 \leq 6$.

Therefore, ρ_2 in Eq. (1.38) is greater than $(2/6^2) \times \sqrt{3}(\sqrt{3} - 1) \simeq 0.035$. Observe that this is much smaller than the value of $\rho = 1/8 = 0.125$ we employed in the continuous approach.

We run the discrete algorithm with the initial data (y_{0h}, y_{1h}) given by the natural approximations of $y_0 = 0$ and y_1 as in Eq. (1.88).

Our first simulations are done with the choice $\rho = 0.035$ for $h = 1/100$. There, the estimated optimal number of iterations is 95 (see Eq. (1.40)), which is much larger than in the continuous approach (where it was 21) due to the fact that ρ is much smaller. In Fig. 1.7 (left), we represent the control $v_h^{k=95,d}$. When compared with the reference control computed for $h = 1/300$ by the discrete method (represented in Fig. 1.10), the relative error $\left\| v_h^{k=95,d} - v_{\text{ref}} \right\|_{L^2} / \left\| v_{\text{ref}} \right\|_{L^2}$ is 5.82%.

In Fig. 1.8, we represent the relative error $\left\| v_h^{k,d} - v_{\text{ref}} \right\|_{L^2} / \left\| v_{\text{ref}} \right\|_{L^2}$ for k between 1 and 50,000. The best iterate is the 54-th one, which corresponds to a relative error of 5.80%. It is represented in Fig. 1.7 (right).

It might seem surprising that the sequence $v_h^{k,d}$ does not converge to v_{ref} as $k \to \infty$. This is actually due to the fact that v_{ref} corresponds to the control computed for $h = 1/300$. Indeed, setting $v_h^{\infty,d}$ the limit of $v_h^{k,d}$ as k goes to infinity, we represent the relative error $\left\| v_h^{k,d} - v_h^{\infty,d} \right\|_{L^2} / \left\| v_h^{\infty,d} \right\|_{L^2}$ in Fig. 1.9. We shall later explain how to compute $v_h^{\infty,d}$, represented in Fig. 1.10 (left).

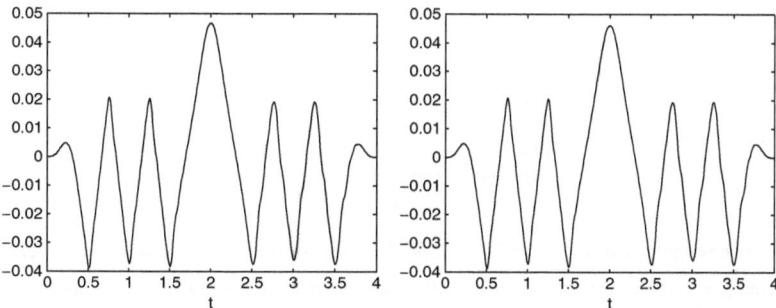

Fig. 1.7 *Left*, the control $v_h^{k=95,d}$ obtained by the discrete approach at the predicted iteration number 95 for $h = 1/100$, $\rho = 0.035$. *Right*, the control $v_h^{k=54,d}$ corresponding to the iterate $k = 54$ that approximates v_{ref} at best.

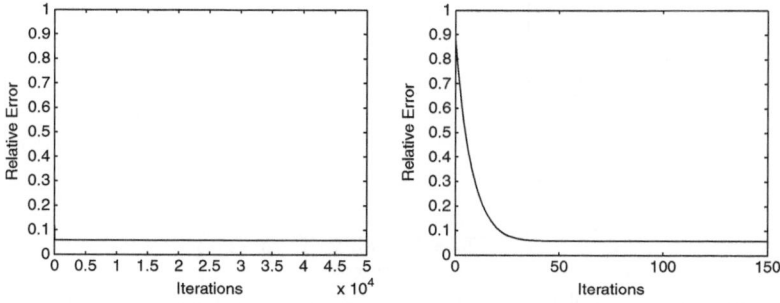

Fig. 1.8 The relative error $\left\| v_h^{k,d} - v_{\text{ref}} \right\|_{L^2} / \left\| v_{\text{ref}} \right\|_{L^2}$ for the discrete approach at each iteration for $h = 1/100$, $\rho = 0.035$. *Left*, for iterations from 0 to 50,000. *Right*, a zoom on the iterations between 0 and 150.

Note that the previous computations are done for $\rho = 0.035$, but as we said, this is only an estimate on the parameter ρ we can choose. In particular, one could also try to take $\rho = 1/8$, which is admissible for the continuous wave equation, though it is a priori out of the valid range of ρ for the semi-discrete equation according to our estimates. For $h = 100$, the estimated iteration is then 42. The corresponding control $v_h^{k=42,d}$ is such that the relative error $\left\| v_h^{k=42,d} - v_{\text{ref}} \right\|_{L^2} / \| v_{\text{ref}} \|_{L^2}$ is of 5.83%. The best iterate is the 14-th one, for which the relative error $\left\| v_h^{k=14,d} - v_{\text{ref}} \right\|_{L^2} / \| v_{\text{ref}} \|_{L^2}$ is of 5.81%. The corresponding plots are very similar to those of the case $\rho = 0.035$. We only plot the relative error $\left\| v_h^{k,d} - v_{\text{ref}} \right\|_{L^2} / \| v_{\text{ref}} \|_{L^2}$ versus the number of iterations in Fig. 1.11.

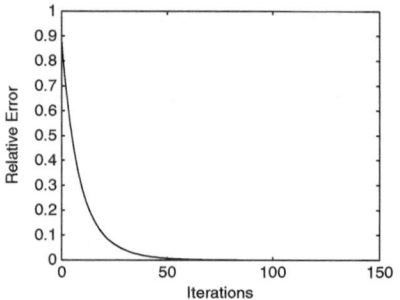

Fig. 1.9 The relative error $\left\| v_h^{k,d} - v_h^{\infty,d} \right\|_{L^2} / \left\| v_h^{\infty,d} \right\|_{L^2}$ for the discrete approach at each iteration for $h = 1/100$, $\rho = 0.035$ and for k from 0 to 150. The relative error is of order 2.10^{-4} at the estimated iteration $k = 95$.

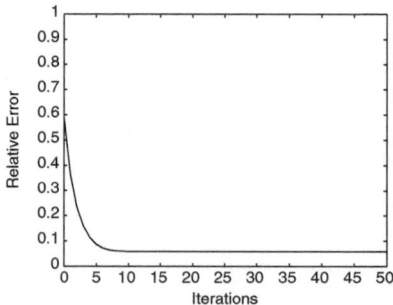

Fig. 1.10 The relative error $\left\| v_h^{k,d} - v_{\text{ref}} \right\|_{L^2} / \| v_{\text{ref}} \|_{L^2}$ for the discrete approach at each iteration for $h = 1/100$, $\rho = 0.125$ for k from 0 to 50.

The Discrete Approach: The Conjugate Gradient Method

In previous paragraphs we underlined the difficulty of estimating the parameters entering into the algorithm. But, as we have explained, in the discrete approach, we also have Eq. (1.44), ensuring the convergence of the minimizer of the functional J_h^{γ} over $\mathscr{V}_h(\gamma/h)$:

$$J_h^{\gamma}(\varphi_{0h}, \varphi_{1h}) = \frac{1}{2} \int_0^T \eta \left| \frac{\varphi_{N,h}}{h} \right|^2 dt - h \sum_{j=1}^N y_{j,0h}\varphi_{j,1h} + h \sum_{j=1}^N y_{j,1h}\varphi_{j,1h}, \qquad (1.93)$$

where φ_h is the solution of Eq. (1.80) with initial data $(\varphi_{0h}, \varphi_{1h}) \in \mathscr{V}_h(\gamma/h)$. In other words, the discrete approach consists in looking for the minimizer of J_h^{γ} over $\mathscr{V}_h(\gamma/h)$.

Since the functional J_h^{γ} is quadratic and well conditioned according to Assumption 3, one can use the conjugate gradient algorithm to compute the minimum $(\Phi_{0h}^d, \Phi_{1h}^d)$ of J_h^{γ} over $\mathscr{V}_h(\gamma/h)$. Doing this, we do not need any estimate on the admissibility and observability constants to run the algorithms. Besides, this algorithm is well known to be much faster than the classical steepest descent one, but with exactly the same complexity.

We therefore run the algorithms for $h = 1/100$ and $h = 1/300$, $\gamma = 1/3$, and the initial data (y_0, y_1) with $y_0 = 0$ and y_1 as in Eq. (1.88). The algorithm converges very fast and it requires only 10 and 9 iterations for $h = 1/100$ and $h = 1/300$, respectively.

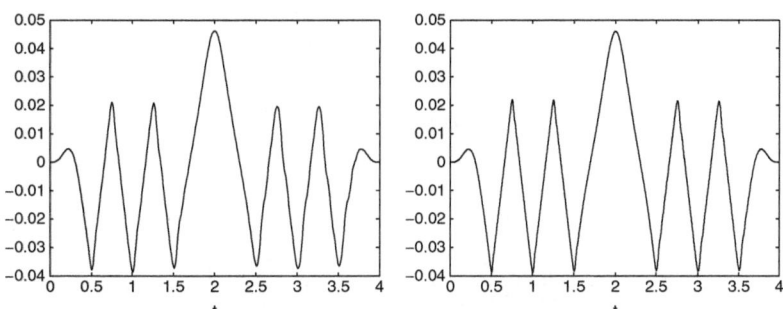

Fig. 1.11 The controls $v_h^{\infty,d}$ for $h = 1/100$ (*left*) and $h = 1/300$ (*right*). We have set $v_{\mathrm{ref}} = v_h^{\infty,d}$ for $h = 1/300$.

In Fig. 1.11, the control $v_h^{k=\infty,d}$ has been computed using the conjugate gradient method as indicated above. The reference control is the one computed for $h = 1/300$.

In Fig. 1.12 we finally represent the rate of convergence of the discrete controls (to be compared with Fig. 1.6). Here again, the slope is -0.97, i.e., much less than -0.67, the slope predicted by our theoretical results in Theorem 1.5. This is again due to the fact that y_1 is more regular than simply $L^2(0,1)$, almost lying in $H^{1/2}(0,1)$; see Remark 1.2.

In higher dimension, there are a few results which prove uniform observability estimates for the wave equation: we refer to [51] for the 2-D case on a uniform mesh, which yields a sharp result. We refer to [41] for the n-dimensional case under general approximation conditions. To our knowledge, the result in [41] is the best one when considering general meshes in any dimension. Still, a precise time estimate for the uniform observability result is missing and whether the filtering scales obtained in [41] are sharp is an open problem.

1.7.2 Distributed Control

System (1) fits in the abstract setting of Eq. (1.1) with $X = H_0^1(\Omega) \times L^2(\Omega)$,

$$A = \begin{pmatrix} 0 & I \\ \Delta & 0 \end{pmatrix}, \quad \mathscr{D}(A) = H^2 \cap H_0^1(\Omega) \times H_0^1(\Omega)$$

and

$$B = \begin{pmatrix} 0 \\ \chi_\omega \end{pmatrix}, \quad U = L^2(\Omega).$$

Indeed, A is skew-adjoint with respect to the scalar product of $X = H_0^1(\Omega) \times L^2(\Omega)$

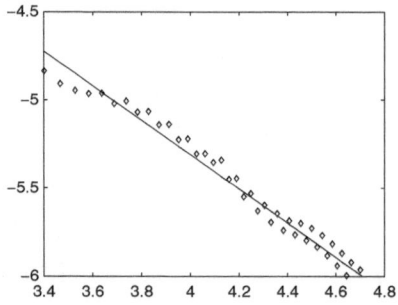

Fig. 1.12 Convergence of the discrete approach: $\left\| v_h^{\infty,d} - v_{\text{ref}} \right\|$ versus h in logarithmic scales. Here, v_{ref} is $v_h^{\infty,d}$ for $h = 1/300$. The plot is done for $h \in (1/120, 1/30)$, the slope being -0.97.

and system (1) is of course admissible since B is bounded from $L^2(\Omega)$ into $L^2(\Omega)$.

Using the scalar product of X, B^* simply reads as $B^* = (0, \chi_\omega)$.

Besides, it is well known that when the GCC (see [3, 5] and the introduction) is satisfied for (ω, Ω, T^*), then the wave equation is observable in time T^*. To be more precise, there exists a constant $C_{\text{obs}} > 0$ such that for all φ solution of

$$\begin{cases} \partial_{tt}\varphi - \Delta\varphi = 0, & (t,x) \in (0,T) \times \Omega, \\ \varphi = 0, & (t,x) \in (0,T) \times \partial\Omega, \\ (\varphi(0,x), \partial_t\varphi(0,x)) = (\varphi_0(x), \varphi_1(x)), \; x \in \Omega, \end{cases} \tag{1.94}$$

we have

$$\|(\varphi_0, \varphi_1)\|^2_{H^1_0(\Omega) \times L^2(\Omega)} \le C^2_{\text{obs}} \int_0^{T^*} \int_\Omega \chi^2_\omega |\partial_t\varphi|^2. \tag{1.95}$$

This is the so-called observability inequality, corresponding to Eq. (1.5) in the abstract setting.

In the following, we assume that Eq. (1.95) holds (or, equivalently, that the Geometric Control Condition holds), and we choose $T > T^*$ and introduce η as in Eq. (1.6).

Note that we made the choice of identifying $H^1_0(\Omega) \times L^2(\Omega)$ with its dual. Doing this, we are thus precisely in the abstract setting of Theorems 1.1, 1.2, and 1.3. However, in applications, one usually identifies $L^2(\Omega)$ with its dual, thus making impossible the identification of $H^1_0(\Omega) \times L^2(\Omega)$ as a reflexive Hilbert space. We shall comment this later on in Remark 1.3.

We are then in position to develop the algorithm in Eqs. (1.17) and (1.18).

1.7.2.1 The Continuous Setting

We divide it in several steps:

Step 0: Set $(\varphi_0^0, \varphi_1^0) = (0,0)$.

The induction formula: Compute φ^k, the solution of

$$\begin{cases} \partial_{tt}\varphi^k - \Delta\varphi^k = 0, & (t,x) \in (0,T) \times \Omega, \\ \varphi^k = 0, & (t,x) \in (0,T) \times \partial\Omega, \\ (\varphi^k(0,x), \partial_t\varphi^k(0,x)) = (\varphi_0^k(x), \varphi_1^k(x)), \; x \in \Omega. \end{cases} \tag{1.96}$$

Then compute ψ^k solution of

$$\begin{cases} \partial_{tt}\psi^k - \Delta\psi^k = -\eta(t)\chi^2_\omega\partial_t\varphi^k, & (t,x) \in (0,T) \times \Omega, \\ \psi^k = 0, & (t,x) \in (0,T) \times \partial\Omega, \\ (\psi^k(T,x), \partial_t\psi^k(T,x)) = (0,0), \; x \in \Omega. \end{cases} \tag{1.97}$$

Finally, set

$$(\varphi_0^{k+1}, \varphi_1^{k+1}) = (\varphi_0^k, \varphi_1^k) - \rho\left((\psi^k(0), \partial_t\psi^k(0)) + (y_0, y_1)\right). \tag{1.98}$$

Note that the map $(\varphi_0^k, \varphi_1^k) \mapsto (\psi^k(0), \partial_t\psi^k(0))$ defined above is precisely the map Λ_T in Eq. (1.13).

Remark 1.3. As we have said, here, we identified $X = H_0^1(\Omega) \times L^2(\Omega)$ with its dual. This allows us to work precisely in the abstract setting of Sect. 1.2.

But our approach also works when identifying $L^2(\Omega)$ with its dual. In that case, we should introduce $X^* = L^2(\Omega) \times H^{-1}(\Omega)$ and, though A is still skew-adjoint with respect to the X-scalar product, we shall introduce A^* the operator defined on $X^* = L^2(\Omega) \times H^{-1}(\Omega)$ by

$$A^* = \begin{pmatrix} 0 & I \\ \Delta & 0 \end{pmatrix}, \quad \mathscr{D}(A^*) = H_0^1(\Omega) \times L^2(\Omega).$$

The duality product between X and X^* is then

$$\left\langle \begin{pmatrix} y_0 \\ y_1 \end{pmatrix}, \begin{pmatrix} \varphi_0 \\ \varphi_1 \end{pmatrix} \right\rangle_{X \times X^*} = \int_\Omega y_1 \varphi_0 - \int_\Omega \nabla y_0 \cdot \nabla (-\Delta)^{-1} \varphi_1.$$

Also, the operator B^* now reads as

$$B^* = (\chi_\omega, 0).$$

The corresponding algorithm then is as follows:

Step 0: Set $(\tilde{\varphi}_0^0, \tilde{\varphi}_1^0) = (0,0)$.

The induction formula: Compute $\tilde{\varphi}^k$, the solution of

$$\begin{cases} \partial_{tt}\tilde{\varphi}^k - \Delta\tilde{\varphi}^k = 0, & (t,x) \in (0,T) \times \Omega, \\ \tilde{\varphi}^k = 0, & (t,x) \in (0,T) \times \partial\Omega, \\ (\tilde{\varphi}^k(0,x), \partial_t\tilde{\varphi}^k(0,x)) = (\tilde{\varphi}_0^k(x), \tilde{\varphi}_1^k(x)), & x \in \Omega. \end{cases} \quad (1.99)$$

Then compute ψ^k solution of

$$\begin{cases} \partial_{tt}\tilde{\psi}^k - \Delta\tilde{\psi}^k = -\eta(t)\chi_\omega^2\tilde{\varphi}^k, & (t,x) \in (0,T) \times \Omega, \\ \tilde{\psi}^k = 0, & (t,x) \in (0,T) \times \partial\Omega, \\ (\tilde{\psi}^k(T,x), \partial_t\tilde{\psi}^k(T,x)) = (0,0), & x \in \Omega. \end{cases} \quad (1.100)$$

Finally set

$$(\tilde{\varphi}_0^{k+1}, \tilde{\varphi}_1^{k+1}) = (\tilde{\varphi}_0^k, \tilde{\varphi}_1^k) - \rho\left(\partial_t\tilde{\psi}^k(0) + y^1, \Delta(\tilde{\psi}^k(0) + y_0)\right). \quad (1.101)$$

Of course, the two algorithms (1.96)–(1.98) and Eqs. (1.99)–(1.101) correspond one to another. Indeed, for all $k \in \mathbb{N}$,

$$\tilde{\varphi}^k = \partial_t\varphi^k, \quad \tilde{\psi}^k = \psi^k,$$

and so for all $k \in \mathbb{N}$, $(\tilde{\varphi}_0^k, \tilde{\varphi}_1^k) = (\varphi_1^k, \Delta\varphi_0^k)$. Hence, of course, the convergence properties of the sequence $(\varphi_0^k, \varphi_1^k)$ proved in Theorem 1.1 have their counterpart for the sequence $(\tilde{\varphi}_0^k, \tilde{\varphi}_1^k)$ (they are basically the same except for a shift in the regularity spaces).

1.7.2.2 The Continuous Approach

Here we introduce the finite-element discretization of the wave equation. The setting we present below is very close to the one in [9] in order to help the readers to see the similarities between the work [9] and our results.

We thus assume that there exists a family $(\tilde{V}_h)_{h>0}$ of finite-dimensional subspaces of $H_0^1(\Omega)$ with the property that there exist $\theta > 0$ and $C > 0$ so that

$$\begin{aligned}
\|(\pi_h\varphi - \varphi)\|_{H_0^1(\Omega)} &\leq Ch^\theta \|\varphi\|_{H^2 \cap H_0^1(\Omega)}, \; \forall \varphi \in H^2 \cap H_0^1(\Omega), \\
\|(\pi_h\varphi - \varphi)\|_{L^2(\Omega)} &\leq Ch^\theta \|\varphi\|_{H_0^1(\Omega)}, \qquad \forall \varphi \in H_0^1(\Omega),
\end{aligned} \tag{1.102}$$

where π_h is the orthogonal projector from $H_0^1(\Omega)$ onto $\tilde{V}_h$.

Note that, on a quasi uniform triangulation $\mathcal{T}_h$, see, e.g., [4], one can take $\theta = 1$ in Eq. (1.102).

We then endow $\tilde{V}_h$ with the $L^2(\Omega)$ scalar product.

We then define the discrete Laplace operator Δ_h as follows:

$$\forall(\varphi_h, \psi_h) \in V_h^2, \qquad \langle -\Delta_h\varphi_h, \psi_h\rangle_{L^2(\Omega)} = \langle \varphi_h, \psi_h\rangle_{H_0^1(\Omega)}.$$

The operator $-\Delta_h$ is then symmetric and positive definite.

We then set B_0 the operator corresponding to the multiplication by the characteristic function χ_ω and set $U_h = B_0^*\tilde{V}_h$, which is of course a subset of $U = L^2(\Omega)$. We then define the operators B_{0h} by $B_{0h}u = \tilde{\pi}_h B_0 u$, where $\tilde{\pi}_h$ is the orthogonal projector of $L^2(\Omega)$ onto $\tilde{V}_h$.

The adjoint of B_{0h} is then given by $B_{0h}^*\varphi = B_0^*\tilde{\pi}_h\varphi$ for $\varphi \in L^2(\Omega)$, which easily implies that the operator norms $\|B_{0h}B_{0h}^*\|_{\mathcal{L}(L^2(\Omega))}$ are uniformly bounded.

To fit into our setting, we thus introduce

$$A_h = \begin{pmatrix} 0 & Id \\ \Delta_h & 0 \end{pmatrix}, \quad B_h = \begin{pmatrix} 0 \\ B_{0h} \end{pmatrix}, \quad V_h = (\tilde{V}_h)^2,$$

with $E_h = Id$ and

$$R_h = \begin{pmatrix} \pi_h & 0 \\ 0 & \pi_h \end{pmatrix}.$$

Assumption 2 immediately follows from the stability of the scheme and the fact that the norms $\|B_{0h}B_{0h}^*\|_{\mathcal{L}(L^2(\Omega))}$ are uniformly bounded.

Then, to prove Assumption 1, we refer to [2, 9]: Assumption 2 holds with θ as in Eq. (1.102) and $s = 2$. Remark that this corresponds to a choice of initial data in $H_{(0)}^3(\Omega) \times H^2 \cap H_0^1(\Omega)$, where $H_{(0)}^3(\Omega)$ is the set of the functions φ of $H^3(\Omega)$ satisfying $\varphi = 0$ and $\Delta\varphi = 0$ on the boundary $\partial\Omega$.

Theorem 1.3 then applies and yields the same convergence results as the one in [9, Theorem 1.1].

To develop the continuous method we need to compute the iteration number K_h^c in Eq. (1.35), and this turns out to be a delicate issue. As explained in Sect. 1.6.1, this requires the knowledge of an approximation of the observability and admissibility constants. Here, the admissibility constant can be taken to be simply T. But evaluating the observability one is a difficult problem.

Very likely, when (ω, Ω, T) satisfies the multiplier condition (requiring that ω is a neighborhood of a part of the boundary Γ of Ω such that $\{x \in \partial\Omega, (x - x^0) \cdot n > 0\} \subset \Gamma$ and $T > 2\sup_\Omega\{|x - x^0|\}$ for some x^0), one can get a reasonable bound on the observability constant. Note however that, even in that case, the observability constant is not explicit since the arguments use a multiplier technique and then a compactness argument, see [31, 36]. Otherwise, if only the GCC is satisfied (see [3]), such bounds on the observability constant are so far unknown.

Let us also emphasize that Assumption 3 does not hold in general, see [16]. This is even the case in $1 - d$ on uniform meshes. However, by suitably filtering the class of initial data, variants of Assumption 3 can be proved. We refer the interested reader to [12, 41, 44, 51] for some nontrivial geometric settings in which Assumption 3 is proved. We shall not develop this point extensively here.

1.8 A Data Assimilation Problem

In this section, we discuss a data assimilation problem that can be treated by the techniques developed in this paper.

1.8.1 The Setting

Under the same notations as before, we consider a system driven by the equation

$$\Phi' = A\Phi, \quad t \geq 0, \qquad \Phi(0) = \Phi_0, \qquad m(t) = B^*\Phi(t). \tag{1.103}$$

We assume that Φ_0 is not known *a priori* but, instead, we have partial measurements on the solution through the measurement $m(t) = B^*\Phi(t)$. The question then is the following: given $m \in L^2(0, T; U)$, can we reconstruct Φ_0?

This problem is of course very much related to the study of the observation map:

$$\mathcal{O} : \begin{cases} X \longrightarrow L^2(0, T; U) \\ \varphi_0 \mapsto B^*\varphi \end{cases}, \tag{1.104}$$

where φ is the solution of Eq. (1.2) with initial data φ_0.

Note that this map $\mathcal{O}$ is well defined in these spaces under the condition (1.3). Also note that the observability inequality (1.5) for (1.2) is completely equivalent to the fact that the map $\mathcal{O}$ has continuous inverse from $L^2(0, T; U) \cap \mathrm{Ran}(\mathcal{O})$ to X.

Therefore, in the following, we will assume the admissibility and observability estimates (1.3)–(1.5) so to guarantee that $\mathcal{O}$ is well defined and invertible on its range.

Of course, this is not enough to obtain an efficient reconstruction algorithm, which is an efficient way to compute the map $\mathcal{O}^{-1}$.

In order to do this, the most natural idea is to introduce the functional

$$\tilde{J}(\varphi_0) = \frac{1}{2} \int_0^T \eta \, \|B^*\varphi - m\|_U^2 \, dt, \tag{1.105}$$

where η is as in Eq. (1.6), or, what is equivalent at the minimization level, since m is assumed to be known,

$$\tilde{J}(\varphi_0) = \frac{1}{2} \int_0^T \eta \, \|B^*\varphi - m\|_U^2 \, dt - \frac{1}{2} \int_0^T \eta \, \|m\|_U^2 \, dt, \tag{1.106}$$

where φ is the solution of Eq. (1.2) with initial data φ_0.

Then $\tilde{J}$ can be rewritten as

$$\tilde{J}(\varphi_0) = \frac{1}{2} \int_0^T \eta \, \|B^*\varphi\|_U^2 \, dt - \int_0^T \eta \, \langle B^*\varphi, m \rangle_U \, dt \tag{1.107}$$

$$= \frac{1}{2} \int_0^T \eta \, \|B^*\varphi\|_U^2 \, dt + \langle \varphi_0, y(0) \rangle_X, \tag{1.108}$$

where $y(0)$ is given by

$$y' = Ay + \eta Bm, \quad t \in (0,T), \quad y(T) = 0. \tag{1.109}$$

Under the form (1.108), the functional $\tilde{J}$ appears as a particular case of the functional J in Eq. (1.9), and therefore, Theorem 1.1 applies.

In order to write our results in a satisfactory way, we only have to check that the degree of smoothness of $y_0 = y(0)$, m, and Φ_0 are all the same.

Indeed, if $\varphi_0 \in \mathscr{D}(A)$, applying Eq. (1.3) and Eq. (1.5) to $A\varphi_0$, we obtain

$$\int_0^{T^*} \left\| B^*\varphi'(t) \right\|_U^2 \, dt \leq C_{ad,T^*}^2 \, \|A\varphi_0\|_X^2,$$

$$\|A\varphi_0\|_X^2 \leq C_{obs,T^*}^2 \int_0^{T^*} \left\| B^*\varphi'(t) \right\|_U^2 \, dt.$$

Therefore, repeating this argument for $\varphi_0 \in \mathscr{D}(A^k)$ and interpolating for $s \geq 0$, we obtain

$$\|B^*\varphi\|_{H^s(0,T^*;U)} \leq C_{ad,T^*} \, \|\varphi_0\|_s, \quad \varphi_0 \in X_s, \tag{1.110}$$

$$\|\varphi_0\|_s \leq C_{obs,T^*} \, \|B^*\varphi\|_{H^s(0,T^*;U)}, \quad \varphi_0 \in X_s; \tag{1.111}$$

These estimates indicate the following fact: for all $s \geq 0$, the map $\mathcal{O}$ maps X_s in $\mathrm{Ran}(\mathcal{O}) \cap H^s(0,T^*;U)$ and has a continuous inverse within these spaces. Equivalently, for all $s \geq 0$, there exists a constant $C_s > 0$ such that

$$\frac{1}{C_s} \|\varphi_0\|_s \leq \|B^*\varphi\|_{H^s(0,T^*;U)} \leq C_s \|\varphi_0\|_s, \quad \varphi_0 \in X_s.$$

Of course, this in particular implies that, if $m \in H^s(0,T;U)$,

$$\|\Phi_0\|_s \leq C_s \|m\|_{H^s(0,T;U)} . \tag{1.112}$$

Let us now explain the fact that, when $m \in H^s(0,T;U)$, $y(0)$ belongs to X_s. If $m \in H^1(0,T;U)$, we differentiate in time the Eq. (1.109) of y:

$$(y')' = A(y') + \eta Bm' + \eta' Bm, \quad t \in (0,T), \qquad y'(T) = 0.$$

Therefore, since B is admissible and $\eta m' + \eta' m \in L^2(0,T;U)$, y' belongs to the space $C([0,T];X)$. Thus, from the Eq. (1.109) of y and the fact that η vanishes at $t = 0$, $Ay(0) = y'(0) \in X$ and then $y(0) \in \mathscr{D}(A)$. This argument can easily be extended to any $s \in \mathbb{N}$ by induction and then to any $s \geq 0$ by interpolation.

We have thus obtained that for all $s \geq 0$, there exists a constant $C_s > 0$ such that

$$\|y(0)\|_s \leq C_s \|m\|_{H^s(0,T;U)} . \tag{1.113}$$

According to this, Theorem 1.1 implies the following:

Theorem 1.8. *Let $s \geq 0$ and $m \in H^s(0,T;U)$. Let $y_0 = y(0)$, where y denotes the solution of Eq. (1.109) and the sequence φ_0^k be defined by Eqs. (1.17) and (1.18).*

Denote by Φ_0 the minimizer of $\tilde{J}$ in Eq. (1.106). Then $\Phi_0 \in X_s$.

Besides, for all $\rho \in (0,\rho_0)$, where ρ_0 is as in Eq. (1.19), the sequence φ_0^k converges to Φ_0 in X and in X_s with the convergence rates (1.23)–(1.24), where δ is given by Eq. (1.20).

Of course, using Eqs. (1.113) and (1.24) implies

$$\left\|\varphi_0^k - \Phi_0\right\|_s \leq C\delta^k (1 + |k|^s) \|m\|_{H^s(0,T;U)}, \quad k \in \mathbb{N}. \tag{1.114}$$

We can then apply the same ideas as the ones used for computing discrete controls.

1.8.2 Numerical Approximation Methods

Let $m_h \in L^2(0,T;U_h)$ and introduce a function $y_{0h} = y_h(0)$, where y_h is the solution of

$$y_h' = A_h y_h + \eta B_h m_h, \quad t \in (0,T), \qquad y_h(T) = 0. \tag{1.115}$$

Then the functionals $\tilde{J}_h$ defined by

$$\tilde{J}_h(\varphi_{0h}) = \frac{1}{2} \int_0^T \eta \|B_h^* \varphi_h - m_h\|_{U_h}^2 \, dt - \frac{1}{2} \int_0^T \eta \|m_h\|_{U_h}^2 \, dt, \tag{1.116}$$

where φ_h is the solution of Eq. (1.11) with initial data φ_{0h}, can be rewritten as follows:

$$\tilde{J}_h(\varphi_{0h}) = \frac{1}{2} \int_0^T \eta \, \|B_h^* \varphi_h\|_{U_h}^2 \, dt + \langle \varphi_{0h}, y_{0h} \rangle_h. \tag{1.117}$$

1.8.2.1 The Continuous Approach

Here, we only suppose that Assumptions 1 and 2 are fulfilled.

Under Assumptions 1 and 2, using Eq. (1.113), Theorem 1.2 applies and yields the following version of Eq. (1.33): for all $k \in \mathbb{N}$,

$$\left\| E_h \varphi_{0h}^k - \varphi_0^k \right\|_X \leq Ck \, \|E_h y_{0h} - y_0\|_X + Ckh^\theta \, \|m\|_{H^s(0,T;U)}. \tag{1.118}$$

Therefore, using Eqs. (1.114) and (1.118) and optimizing in k, setting K_h^c as in Eq. (1.35), we obtain, for some constant independent of h,

$$\left\| E_h \varphi_{0h}^{K_h^c} - \Phi_0 \right\|_X \leq C |\log h|^{\max\{1,s\}} h^\theta \, \|m\|_{H^s(0,T;U)}$$
$$+ C |\log h| \, \|E_h y_{0h} - y_0\|_X. \tag{1.119}$$

In particular, if $\|E_h y_{0h} - y_0\|_X$ tends to zero as $h \to 0$ faster than $1/|\log(h)|$, we have a convergence estimate for this data assimilation problem. Of course, a discrete sequence y_{0h} such that $E_h y_{0h}$ converges to y_0 in X can be built by assuming suitable convergence assumptions of m_h towards m and the convergence of the numerical scheme (1.115) towards the continuous equation (1.109).

Note that it can be necessary to consider the regularity of the measurement $m = B^* \Phi$ in the space variable. Let us give a precise example corresponding to the case of distributed observation, see Sect. 1.7.2, corresponding to

$$B^* = \begin{pmatrix} 0 & 0 \\ 0 & \chi_\omega \end{pmatrix}$$

on $X = H_0^1(\Omega) \times L^2(\Omega)$. There, $B = B^*$ and U can be taken to coincide with X. If furthermore the function χ_ω that localizes the effect of the control in ω is smooth, B (and thus B^*) maps X_s to itself for any $s \geq 0$ (these assumptions are very close to the ones in [9, 25] on the control/observation operator). Therefore, in that case, if $\Phi_0 \in X_k$ ($k \in \mathbb{N}$), $m = B^* \Phi$ belongs to $C^k([0,T];X) \cap C^0([0,T];X_k)$. Note that the $H^k(0,T;X)$-norm of m is then equivalent to its $C^k([0,T];X) \cap C^0([0,T];X_k)$-norm by Eq. (1.112) together with classical energy estimates for solutions of Eq. (1.103). Therefore, a natural space for the measurement m would rather be $C^s([0,T];X) \cap C^0([0,T];X_s)$ and one could therefore simply take the approximate measurement $m_h = R_h m$.

The obtained algorithm is actually very close to the one derived in [25] from the continuous "algorithm" in [29] and suffers from the same disadvantages and in particular from the difficulty of computing the stopping time.

1.8.2.2 The Discrete Approach

In this paragraph, we suppose that Assumptions 1, 2, and 3 hold.

Using Theorem 1.4 and Eq. (1.113), one can obtain the following version of Eq. (1.39): for all $k \in \mathbb{N}$,

$$\left\| E_h \varphi_{0h}^k - \varphi_0^k \right\|_X \le k\rho \left(\| E_h y_{0h} - y_0 \|_X + Ch^\theta \| m \|_{H^s(0,T;U)} \right). \tag{1.120}$$

In particular, based on this estimate and Eq. (1.114), we obtain that for some constant C independent of k and h, for all $k \ge K_h^d$ [given by Eq. (1.40)],

$$\left\| E_h \varphi_{0h}^k - \Phi_0 \right\|_X \le Ch^\theta \| m \|_{H^s(0,T;U)} + C \| E_h y_{0h} - y_0 \|_X. \tag{1.121}$$

In particular, similarly as in Eq. (1.44),

$$\| E_h \Phi_{0h} - \Phi_0 \|_X \le Ch^\theta \| m \|_{H^s(0,T;U)} + C \| E_h y_{0h} - y_0 \|_X, \tag{1.122}$$

where Φ_{0h} is the minimizer of $\tilde{J}_h$ in Eq. (1.116).

Remark also that, similarly as in Sect. 1.6.2, if one can guarantee that y_0 given by Eq. (1.109) and y_{0h} given by Eq. (1.115) are such that $E_h y_{0h}$ strongly converge in X to y_0, one can guarantee that $E_h \Phi_{0h}$ strongly converge to Φ_0. Such convergences for the sequence $E_h y_{0h}$ are very natural for sequences of observations m_h that strongly converge to m in $L^2(0,T;U)$ (this statement has to be made more precise by explaining how $m_h \in L^2(0,T;U_h)$ is identified as an element of $L^2(0,T;U)$).

Of course, this implies that, similarly as in Sect. 1.6.3, $\tilde{J}_h$ can be minimized using faster algorithms than the steepest descent one and in particular the conjugate gradient method.

Chapter 2
Observability for the $1-d$ Finite-Difference Wave Equation

2.1 Objectives

In this chapter, we discuss the observability properties for the $1-d$ finite-difference wave equation.

For the convenience of the reader, let us recall the equations, already introduced in Eq. (1.80).

Let $N \in \mathbb{N}$, $h = 1/(N+1)$. Given $(\varphi_{0h}, \varphi_{1h})$, compute the solution φ_h of the following system:

$$\begin{cases} \partial_{tt} \varphi_{j,h} - \dfrac{1}{h^2} \left(\varphi_{j+1,h} - 2\varphi_{j,h} + \varphi_{j-1,h} \right) = 0, & (t,j) \in (0,T) \times \{1, \cdots, N\}, \\ \varphi_{0,h}(t) = \varphi_{N+1,h}(t) = 0, & t \in (0,T), \\ (\varphi_h(0), \partial_t \varphi_h(0)) = (\varphi_{0h}, \varphi_{1h}). \end{cases} \tag{2.1}$$

Here, we will not be interested in any convergence process, but rather try to prove some estimates uniformly with respect to $h > 0$, and in particular uniform admissibility and observability results. Before going further, let us also emphasize that this uniform admissibility result will be an important step in the proof of the convergence of the discrete waves towards the continuous ones when working with boundary data in $L^2(0,T)$.

Note that the discrete equation (2.1), as its continuous counterpart, is conservative in the sense that its energy

$$E_h[\varphi_h](t) = h \sum_{j=1}^{N} |\partial_t \varphi_j(t)|^2 + h \sum_{j=0}^{N} \left(\frac{\varphi_{j+1}(t) - \varphi_j(t)}{h} \right)^2, \tag{2.2}$$

sometimes simply denoted by $E_h(t)$ when no confusion may occur, is constant in time:

$$\forall t \geq 0, \quad E_h(t) = E_h(0). \tag{2.3}$$

S. Ervedoza and E. Zuazua, *Numerical Approximation of Exact Controls for Waves*, SpringerBriefs in Mathematics, DOI 10.1007/978-1-4614-5808-1_2, © Sylvain Ervedoza and Enrique Zuazua 2013

2.2 Spectral Decomposition of the Discrete Laplacian

In this section, we briefly recall the spectral decomposition of the discrete Laplacian.

To be more precise, we consider the eigenvalue problem associated with the 3-point finite-difference scheme for the $1-d$ Laplacian:

$$\begin{cases} -\dfrac{w_{j+1}+w_{j-1}-2w_j}{h^2} = \lambda w_j, & j=0,\cdots,N+1, \\ w_0 = w_{N+1} = 0. \end{cases} \tag{2.4}$$

A simple iteration process shows that if $w_1 = 0$ and w solves (2.4), then $w_j = 0$ for all $j \in \{0,\cdots,N+1\}$. Hence all the eigenvalues are simple.

Furthermore the spectrum of the discrete Laplacian is given by the sequence of eigenvalues

$$0 < \lambda_1(h) < \lambda_2(h) < \cdots < \lambda_N(h),$$

which can be computed explicitly

$$\lambda_k(h) = \frac{4}{h^2}\sin^2\left(\frac{\pi k h}{2}\right), \quad k=1,\cdots,N. \tag{2.5}$$

The eigenvector $w^k = \left(w_1^k,\cdots,w_N^k\right)$ associated to the eigenvalue $\lambda_k(h)$ can also be computed explicitly:

$$w_j^k = \sqrt{2}\sin\left(\pi k j h\right), \quad j=1,\cdots,N. \tag{2.6}$$

Observe in particular that the eigenvectors of the discrete system do not depend on $h > 0$ and coincide with the restriction of the continuous eigenfunctions $w^k(x) = \sqrt{2}\sin(k\pi x)$ of the Laplace operator on $(0,1)$ to the discrete mesh.

Let us now compare the eigenvalues of the discrete Laplace operator Δ_h and the continuous one ∂_{xx}:

- For fixed k, $\lim_{h\to 0}\lambda_k(h) = \pi^2 k^2$, which is the k-th eigenvalue of the continuous Laplace operator $-\partial_{xx}$ on $(0,1)$.
- We have the following bounds:

$$\frac{4}{\pi^2}k^2\pi^2 \le \lambda_k(h) \le k^2\pi^2 \quad \text{for all} \quad 0 < h < 1, \quad 1 \le k \le N. \tag{2.7}$$

- The discrete eigenvalues $\sqrt{\lambda_k(h)}$ uniformly converge to the corresponding continuous ones $k\pi$ when $k = o(1/h^{2/3})$ since, at first order,

$$\left|\sqrt{\lambda_k(h)} - k\pi\right| \sim Ck^3 h^2. \tag{2.8}$$

Let us now recall some orthogonality properties of the eigenvectors that can be found, e.g., in [28]:

Lemma 2.1. *For any eigenvector w with eigenvalue λ of Eq. (2.4) the following identity holds:*

$$h \sum_{j=0}^{N} \left| \frac{w_{j+1} - w_j}{h} \right|^2 = \lambda h \sum_{j=1}^{N} |w_j|^2. \tag{2.9}$$

The eigenvectors $(w^k)_{k \in \{1, \cdots, N\}}$ in (2.6) satisfy:

$$h \sum_{j=1}^{N} w_j^k w_j^\ell = \delta_{k\ell}, \tag{2.10}$$

and

$$h \sum_{j=0}^{N} \left(\frac{w_{j+1}^k - w_j^k}{h} \right) \left(\frac{w_{j+1}^\ell - w_j^\ell}{h} \right) = \lambda_k \delta_{k\ell}, \tag{2.11}$$

where $\delta_{k\ell}$ is the Kronecker symbol.

2.3 Uniform Admissibility of Discrete Waves

For convenience and later use, we begin by stating a uniform admissibility result, which can also be found in [28] and will be useful for studying the convergence of the discrete normal derivatives of the solutions of Eq. (2.1) towards the continuous ones.

Theorem 2.1. *For all time $T > 0$ there exists a finite positive constant $C(T) > 0$ such that*

$$\int_0^T \left| \frac{\varphi_N(t)}{h} \right|^2 dt \le C(T) E_h(0), \tag{2.12}$$

for all solution φ_h of the adjoint equation (2.1) and for all $h > 0$. Besides, we can take $C(T) = T + 2$.

The proof of Theorem 2.1 is briefly given in Sect. 2.3.2. It is based on a multiplier identity given in the next section.

2.3.1 The Multiplier Identity

Our results are based on the following multiplier identity that can be found in [28]:

Theorem 2.2. *For all $h > 0$ and $T > 0$ any solution φ_h of Eq. (2.1) satisfies*

$$TE_h(0) + X_h(t) \Big|_0^T = \int_0^T \left| \frac{\varphi_N(t)}{h} \right|^2 dt + \frac{h^3}{2} \sum_{j=0}^{N} \int_0^T \left| \frac{\partial_t \varphi_{j+1} - \partial_t \varphi_j}{h} \right|^2 dt, \tag{2.13}$$

with

$$X_h(t) = 2h \sum_{j=1}^{N} jh \left(\frac{\varphi_{j+1} - \varphi_{j-1}}{2h} \right) \partial_t \varphi_j. \tag{2.14}$$

The proof of Theorem 2.2 uses the multiplier $j(\varphi_{j+1} - \varphi_{j-1})$, which is the discrete counterpart of $x \partial_x \varphi$. Integrating by parts in space (in a discrete manner) and time, we obtain (2.13). We refer to [28] for the details of the computations. We only sketch it below since it will be useful later on in Chap. 4.

Proof (Sketch). Multiplying the Eq. (2.1) by $jh(\varphi_{j+1} - \varphi_{j-1})/h$, we have

$$h \sum_{j=1}^{N} \int_0^T \partial_{tt} \varphi_j \, jh \left(\frac{\varphi_{j+1} - \varphi_{j-1}}{h} \right) dt$$

$$= h \sum_{j=1}^{N} \int_0^T \frac{1}{h^2} (\varphi_{j+1} - 2\varphi_j + \varphi_{j-1}) \, jh \left(\frac{\varphi_{j+1} - \varphi_{j-1}}{h} \right) dt.$$

After tedious computations, one shows (cf. [28])

$$h \sum_{j=1}^{N} \int_0^T \partial_{tt} \varphi_j \, jh \left(\frac{\varphi_{j+1} - \varphi_{j-1}}{h} \right) dt = X_h(t) \Big|_0^T + h \sum_{j=1}^{N} \int_0^T |\partial_t \varphi_j|^2 \, dt$$

$$- \frac{h^3}{2} \sum_{j=0}^{N} \int_0^T \left| \frac{\partial_t \varphi_{j+1} - \partial_t \varphi_j}{h} \right|^2 dt$$

and

$$h \sum_{j=1}^{N} \int_0^T \frac{1}{h^2} (\varphi_{j+1} - 2\varphi_j + \varphi_{j-1}) \, jh \left(\frac{\varphi_{j+1} - \varphi_{j-1}}{h} \right) dt$$

$$= \int_0^T \left| \frac{\varphi_N(t)}{h} \right|^2 dt - h \sum_{j=0}^{N} \int_0^T \left(\frac{\varphi_{j+1} - \varphi_j}{h} \right)^2 dt.$$

Putting these identities together yields (2.13).   □

2.3.2 Proof of the Uniform Hidden Regularity Result

Proof (Theorem 2.1). This is an immediate consequence of Theorem 2.2. It suffices to bound the time boundary terms $X_h(T) - X_h(0)$ by the energy E_h to get the result:

$$|X_h| \leq 2 \left[h \sum_{j=1}^{N} |\partial_t \varphi_j|^2 \right]^{1/2} \left[h \sum_{j=1}^{N} \left| jh \left(\frac{\varphi_{j+1} - \varphi_{j-1}}{2h} \right) \right|^2 \right]^{1/2}$$

$$\leq 2 \left[h \sum_{j=1}^{N} |\partial_t \varphi_j|^2 \right]^{1/2} \left[h \sum_{j=1}^{N} \left(\frac{\varphi_{j+1} - \varphi_{j-1}}{2h} \right)^2 \right]^{1/2} \leq E_h. \qquad (2.15)$$

This concludes the proof of Theorem 2.1. □

2.4 An Observability Result

The goal of this section is to show the following result:

Theorem 2.3. *Assume that $\gamma < 1$. Then for all T such that*

$$T > T(\gamma) = 2/\cos(\pi\gamma/2), \qquad (2.16)$$

for every solution φ_h of Eq. (1.80) in the class

$$\mathscr{V}_h(\gamma/h) = Span\left\{ w^k, \quad kh \leq \gamma \right\}$$

uniformly as $h \to 0$, we have

$$\left(T \cos^2 \left(\frac{\gamma\pi}{2} \right) - 2\cos \left(\frac{\pi\gamma}{2} \right) - \frac{h}{2} \right) E_h(0) \leq \int_0^T \left| \frac{\varphi_N}{h} \right|^2 dt, \qquad (2.17)$$

where E_h is the discrete energy of solutions of Eq. (2.1) defined in Eq. (2.2).

The proof of Theorem 2.3 is based on the discrete multiplier identity in Theorem 2.2 (and developed in [28]). However, the estimates we explain below yield a sharp result on the uniform time of observability for discrete waves with an explicit uniform observability constant, thus improving the estimates in [28].

2.4.1 Equipartition of the Energy

We also recall the following proof of the so-called property of equipartition of the energy for discrete waves:

Lemma 2.2 (Equipartition of the energy). *For $h > 0$ and φ_h solution of Eq. (2.1),*

$$-h \sum_{j=1}^{N} \int_0^T |\partial_{tt} \varphi_j|^2 dt + h \sum_{j=0}^{N} \int_0^T \left| \frac{\partial_t \varphi_{j+1} - \partial_t \varphi_j}{h} \right|^2 dt + Y_h(t) \Big|_0^T = 0, \qquad (2.18)$$

where

$$Y_h(t) = h \sum_{j=1}^{N} \partial_{tt} \varphi_j \partial_t \varphi_j. \tag{2.19}$$

Again, for the proof of Lemma 2.2, we refer to [28].

2.4.2 The Multiplier Identity Revisited

From now on, we do not follow anymore the proofs of [28] but rather try to optimize them to improve the obtained estimates.

We introduce a modified energy $\tilde{E}_h$ for solutions φ_h or Eq. (2.1). First, remark that any φ_h solution of Eq. (2.1) can be developed on the basis of eigenfunctions of $-\Delta_h$ as follows:

$$\varphi_h(t) = \sum_{|k| \leq N} \hat{\varphi}_k e^{i\mu_k(h)t} w^{|k|} \tag{2.20}$$

with $\mu_k(h) = \sqrt{\lambda_k(h)}$ for $k > 0$ and $\mu_{-k}(h) = -\mu_k(h)$.

According to Lemma 2.1, its energy reads as

$$E_h[\varphi_h] = 2 \sum_{|k| \leq N} |\hat{\varphi}_k|^2 \lambda_k(h). \tag{2.21}$$

Similarly, the energy of $\partial_t \varphi_h$, which is also a solution of Eq. (2.1), and that we shall denote by $E_h[\partial_t \varphi_h]$ to avoid confusion, can be rewritten as

$$E_h[\partial_t \varphi_h] = 2 \sum_{|k| \leq N} |\hat{\varphi}_k|^2 \lambda_k(h)^2.$$

Note that, of course, $E_h[\varphi_h]$ and $E_h[\partial_t \varphi_h]$ are independent of time since φ_h and $\partial_t \varphi_h$ are solutions of Eq. (2.1).

We then introduce

$$\tilde{E}_h[\varphi_h] = E_h[\varphi_h] - \frac{h^2}{4} E_h[\partial_t \varphi_h]. \tag{2.22}$$

This modified energy is thus constant in time and satisfies

$$\tilde{E}_h[\varphi_h] = 2 \sum_{|k| \leq N} |\hat{\varphi}_k|^2 \lambda_{|k|}(h) \cos^2 \left(\frac{k\pi h}{2} \right). \tag{2.23}$$

We are now in position to state the following multiplier identity:

Theorem 2.4. *For all $h > 0$ and $T > 0$, any solution φ_h of Eq. (2.1) satisfies*

$$T\tilde{E}_h[\varphi_h] + Z_h(t) \Big|_0^T = \int_0^T \left| \frac{\varphi_N(t)}{h} \right|^2 dt \tag{2.24}$$

with

$$Z_h(t) = X_h(t) + \frac{h^2}{4}Y_h(t), \quad \text{with } Y_h(t) = h\sum_{j=1}^{N} \partial_t \varphi_j \partial_{tt} \varphi_j. \quad (2.25)$$

Proof. To simplify the notations, we do not make explicit the dependence in $h > 0$, which is assumed to be fixed along the computations.

According to Lemma 2.2, since $\partial_t \varphi_h$ is a solution of Eq. (2.1), the following identity holds:

$$h\sum_{j=0}^{N} \int_0^T \left| \frac{\partial_t \varphi_{j+1} - \partial_t \varphi_j}{h} \right|^2 dt = \frac{h}{2}\sum_{j=1}^{N} \int_0^T |\partial_{tt}\varphi_j|^2 dt$$

$$+ \frac{h}{2}\sum_{j=0}^{N} \int_0^T \left| \frac{\partial_t \varphi_{j+1} - \partial_t \varphi_j}{h} \right|^2 dt - \frac{Y_h(t)}{2}\Big|_0^T, \quad (2.26)$$

where Y_h is as in Eq. (2.25).

Of course,

$$h\sum_{j=1}^{N} \int_0^T |\partial_t \varphi_j|^2 dt + h\sum_{j=0}^{N} \int_0^T \left| \frac{\partial_t \varphi_{j+1} - \partial_t \varphi_j}{h} \right|^2 dt = TE_h[\partial_t \varphi_h],$$

and then Eq. (2.24) follows from Eqs. (2.26) and (2.13). □

2.4.3 Uniform Observability for Filtered Solutions

We now focus on the proof of Theorem 2.3. It mainly consists in estimating the terms in Eq. (2.24) and in particular $Z_h(t)$.

2.4.3.1 Estimates on $Y_h(t)$

Let us begin with the following bound on Y_h:

Lemma 2.3. *For all $h > 0$ and $t \geq 0$, for any solution φ_h of Eq. (2.1),*

$$h^2|Y_h(t)| \leq hE_h[\varphi_h]. \quad (2.27)$$

Proof. Computing $h^2 Y_h$ we get

$$h^2 Y_h(t) = h\sum_{j=1}^{N} \partial_t \varphi_j (h^2 \partial_{tt} \varphi_j)$$

$$= h\sum_{j=1}^{N} \partial_t \varphi_j (\varphi_{j+1} - 2\varphi_j + \varphi_{j-1})$$

$$= h^2 \sum_{j=1}^{N} \partial_t \varphi_j \left(\frac{\varphi_{j+1} - \varphi_j}{h} \right) - h^2 \sum_{j=1}^{N} \partial_t \varphi_j \left(\frac{\varphi_j - \varphi_{j-1}}{h} \right).$$

But

$$2 \left| h \sum_{j=1}^{N} \partial_t \varphi_j \left(\frac{\varphi_{j+1} - \varphi_j}{h} \right) \right| \leq E_h(t),$$

and thus estimate (2.27) follows immediately.   □

2.4.3.2 Estimates on $X_h(t)$

This is the most technical step of our proof. The idea is to use the Fourier decomposition of solutions φ_h of Eq. (2.1) to bound X_h conveniently.

Proposition 2.1. *For all $h > 0$, $t \geq 0$, and $\gamma \in (0,1)$, any solution φ_h of Eq. (2.1) with data in $\mathscr{V}_h(\gamma/h)$ satisfies*

$$|X_h(t)| \leq \frac{\tilde{E}_h[\varphi_h]}{\cos\left(\dfrac{\gamma\pi}{2} \right)}. \tag{2.28}$$

Proof. Let us begin by computing $\tilde{E}_h[\varphi_h]$ at some time t, for instance, $t = 0$, in terms of the Fourier coefficients of $\varphi_h(t), \partial_t \varphi_h(t)$. If

$$\varphi_h^0 = \sum_{k=1}^{N} \hat{a}_k w^k, \quad \varphi_h^1 = \sum_{\ell=1}^{N} \hat{b}_\ell w^\ell,$$

then $\tilde{E}_h$ can be written as

$$\tilde{E}_h = \sum_{k=1}^{N} |\hat{a}_k|^2 \lambda_k(h) \cos^2\left(\frac{k\pi h}{2} \right) + \sum_{\ell=1}^{N} |\hat{b}_\ell|^2 \cos^2\left(\frac{\ell\pi h}{2} \right). \tag{2.29}$$

Proposition 2.1 is then a direct consequence of the following lemma:

Lemma 2.4. *Let a_h and b_h be two discrete functions which can be written as*

$$a_h = \sum_{k=1}^{N} \hat{a}_k w^k, \quad b_h = \sum_{\ell=1}^{N} \hat{b}_\ell w^\ell.$$

Then, setting

$$X_h(a_h, b_h) = 2h \sum_{j=1}^{N} jh \left(\frac{a_{j+1} - a_{j-1}}{2h} \right) b_j,$$

we have

$$|X_h(a_h, b_h)| \leq 2 \left(\sum_{k=1}^{N} |\hat{a}_k|^2 \lambda_k(h) \cos^2\left(\frac{k\pi h}{2} \right) \right)^{1/2} \left(\sum_{\ell=1}^{N} |\hat{b}_\ell|^2 \right)^{1/2}. \tag{2.30}$$

In particular, if we assume that, for some $\gamma \in (0,1)$,

$$\hat{a}_k = \hat{b}_\ell = 0, \quad \forall k, \ell \geq \gamma(N+1), \tag{2.31}$$

then

$$|X_h(a_h, b_h)| \leq \frac{1}{\cos\left(\frac{\gamma\pi}{2}\right)} \left[\sum_{k=1}^{N} |\hat{a}_k|^2 \lambda_k(h) \cos^2\left(\frac{k\pi h}{2}\right) \right.$$

$$\left. + \sum_{\ell=1}^{N} |\hat{b}_\ell|^2 \cos^2\left(\frac{\ell\pi h}{2}\right) \right]. \tag{2.32}$$

Of course, Lemma 2.4 and in particular estimate (2.32), proved hereafter, immediately yield (2.28). □

Proof (Lemma 2.4). For all $j \in \{1, \cdots, N\}$,

$$\frac{a_{j+1} - a_{j-1}}{2h} = \sqrt{2} \sum_{k=1}^{N} \hat{a}_k \cos(k\pi jh) \frac{\sin(k\pi h)}{h}.$$

Thus,

$$X_h(a_h, b_h) = 4h \sum_{j=1}^{N} jh \left(\sum_{k=1}^{N} \hat{a}_k \cos(k\pi jh) \frac{\sin(k\pi h)}{h} \right) \left(\sum_{\ell=1}^{N} \hat{b}_\ell \sin(\ell\pi jh) \right).$$

Therefore, by orthogonality properties of the discrete cosine functions (the counterpart of Lemma 2.1 with the cosine functions),

$$|X_h(a_h, b_h)|^2$$

$$\leq 4 \left(2h \sum_{j=1}^{N} \left(\sum_{k=1}^{N} \hat{a}_k \cos(k\pi jh) \frac{\sin(k\pi h)}{h} \right)^2 \right) \left(2h \sum_{j=1}^{N} \left(\sum_{\ell=1}^{N} \hat{b}_\ell \sin(\ell\pi jh) \right)^2 \right)$$

$$\leq 4 \left(\sum_{k=1}^{N} |\hat{a}_k|^2 \left(\frac{\sin(k\pi h)}{h} \right)^2 \right) \left(\sum_{\ell=1}^{N} |\hat{b}_\ell|^2 \right),$$

where we used that, for all sequence $(\alpha_k)_{1 \leq k \leq N}$,

$$2h \sum_{j=1}^{N} \left(\sum_{k=1}^{N} \alpha_k \cos(k\pi jh) \right)^2 = \sum_{k=1}^{N} |\alpha_k|^2.$$

Note then that

$$\left(\frac{\sin(k\pi h)}{h} \right)^2 = \frac{4}{h^2} \sin^2\left(\frac{k\pi h}{2}\right) \cos^2\left(\frac{k\pi h}{2}\right) = \lambda_k(h) \cos^2\left(\frac{k\pi h}{2}\right).$$

The bound (2.30) immediately follows.

If we assume (2.31), then by Cauchy–Schwarz inequality, Eq. (2.30) implies

$$|X_h(a_h, b_h)| \leq \frac{1}{\cos\left(\frac{\gamma\pi}{2}\right)} \sum_{k=1}^{N} |\hat{a}_k|^2 \lambda_k(h) \cos^2\left(\frac{k\pi h}{2}\right) + \cos\left(\frac{\gamma\pi}{2}\right) \sum_{\ell=1}^{N} |\hat{b}_\ell|^2,$$

and the last term satisfies:

$$\cos\left(\frac{\gamma\pi}{2}\right) \sum_{\ell=1}^{N} |\hat{b}_\ell|^2 \leq \frac{1}{\cos\left(\frac{\gamma\pi}{2}\right)} \sum_{\ell=1}^{N} |\hat{b}_\ell|^2 \cos^2\left(\frac{\ell\pi h}{2}\right),$$

and estimate (2.32) follows immediately. $\square$

2.4.4 Proof of Theorem 2.3

Proof (Theorem 2.3). Identity (2.24) and estimates (2.27) and (2.28) imply that any solution φ_h of Eq. (2.1) in the class $\mathcal{V}_h(\gamma/h)$ satisfies

$$\left| T\tilde{E}_h[\varphi_h] - \int_0^T \left|\frac{\varphi_N}{h}\right|^2 dt \right| \leq \frac{2}{\cos\left(\frac{\gamma\pi}{2}\right)} \tilde{E}_h(\varphi_h) + \frac{h}{2} E_h(\varphi_h). \qquad (2.33)$$

Therefore,

$$\left(T - \frac{2}{\cos\left(\frac{\gamma\pi}{2}\right)} \right) \tilde{E}_h[\varphi_h] - \frac{h}{2} E_h[\varphi_h] \leq \int_0^T \left|\frac{\varphi_N}{h}\right|^2 dt.$$

But, since φ_h belongs to the class $\mathcal{V}_h(\gamma/h)$, the Fourier expressions of the energy $E_h[\varphi_h]$ in Eq. (2.21) and $\tilde{E}_h[\varphi_h]$ in Eq. (2.23) yield

$$\cos\left(\frac{\gamma\pi}{2}\right)^2 E_h[\varphi_h] \leq \tilde{E}_h[\varphi_h], \qquad (2.34)$$

which concludes the proof of Theorem 2.3. $\square$

Chapter 3
Convergence of the Finite-Difference Method for the $1-d$ Wave Equation with Homogeneous Dirichlet Boundary Conditions

3.1 Objectives

This chapter of the book is devoted to the study of the convergence of the numerical scheme

$$\begin{cases} \partial_{tt}\varphi_{j,h} - \dfrac{1}{h^2}\left(\varphi_{j+1,h} - 2\varphi_{j,h} + \varphi_{j-1,h}\right) = 0, \\ \qquad\qquad\qquad\qquad (t,j) \in (0,T) \times \{1,\dots,N\}, \\ \varphi_{0,h}(t) = \varphi_{N+1,h}(t) = 0, \qquad\qquad\quad t \in (0,T), \\ (\varphi_h(0), \partial_t\varphi_h(0)) = (\varphi_{0h}, \varphi_{1h}), \end{cases} \qquad (3.1)$$

towards the continuous wave equation

$$\begin{cases} \partial_{tt}\varphi - \partial_{xx}\varphi = 0, & (t,x) \in (0,T) \times (0,1), \\ \varphi(t,0) = \varphi(t,1) = 0, & t \in (0,T), \\ (\varphi(0), \partial_t\varphi(0)) = (\varphi_0, \varphi_1). \end{cases} \qquad (3.2)$$

Of course, first of all, one needs to explain how discrete and continuous solutions can be compared. This will be done in Sect. 3.2. In Sect. 3.3, we will present our main convergence result. We shall then present some further convergence results in Sect. 3.4 and illustrate them in Sect. 3.5.

3.2 Extension Operators

We first describe the extension operators we shall use. We will then explain how the obtained results can be interpreted in terms of the more classical extension operators.

S. Ervedoza and E. Zuazua, *Numerical Approximation of Exact Controls for Waves*,   59
SpringerBriefs in Mathematics, DOI 10.1007/978-1-4614-5808-1_3,
© Sylvain Ervedoza and Enrique Zuazua 2013

3.2.1 The Fourier Extension

For $h > 0$, given a discrete function $a_h = (a_{j,h})_{j \in \{1,\ldots,N\}}$ (with $N + 1 = 1/h$), since the sequence w_h^k is an orthonormal basis for the $h \langle \cdot, \cdot \rangle_{\ell^2(\mathbb{R}^N)}$-norm due to Lemma 2.1, there exist coefficients $\hat{a}_k$ such that

$$a_h = \sum_{k=1}^N \hat{a}_k w_h^k, \quad [\text{recall that } w_{j,h}^k = \sqrt{2}\sin(k\pi jh)] \tag{3.3}$$

in the sense that, for all $j \in \{1,\ldots,N\}$,

$$a_{j,h} = \sum_{k=1}^N \hat{a}_k \sqrt{2}\sin(k\pi jh). \tag{3.4}$$

Of course, this yields a natural Fourier extension denoted by $\mathbb{F}_h$ for discrete functions a_h given by Eq. (3.3):

$$\mathbb{F}_h(a_h)(x) = \sum_{k=1}^N \hat{a}_k \sqrt{2}\sin(k\pi x), \quad x \in (0,1). \tag{3.5}$$

The advantage of this definition is that now $\mathbb{F}_h(a_h)$ is a smooth function of x.

The energy of a solution φ_h of Eq. (3.1) at time t, given by Eq. (2.2), is then equivalent, uniformly with respect to $h > 0$, to the $H_0^1(0,1) \times L^2(0,1)$-norm of $(\mathbb{F}_h(\varphi_h), \mathbb{F}_h(\varphi_h'))$. This issue will be discussed in Proposition 3.3 below.

Another interesting feature of this Fourier extension is that, due to the discrete orthogonality properties of the eigenvectors w^k proved in Lemma 2.1 and their usual $L^2(0,1)$-orthogonality, i.e., $\int_0^1 w^k(x)w^\ell(x)\,dx = \delta_{k,\ell}$ for all $k, \ell \in \mathbb{N}$, for all discrete functions a_h, b_h, we have

$$h \sum_{j=1}^N a_{j,h} b_{j,h} = \int_0^1 \mathbb{F}_h(a_h)\mathbb{F}_h(b_h)\,dx.$$

This fact will be used to simplify some expressions.

3.2.2 Other Extension Operators

When using finite-difference (or finite element) methods, the Fourier extension is not the most natural one. Given a discrete function $a_h = (a_{j,h})_{j \in \{1,\ldots,N\}}$ (with $N + 1 = 1/h$), consider the classical extension operators $\mathbb{P}_h$ and $\mathbb{Q}_h$ defined by

$$\mathbb{P}_h(a_h)(x) = a_{j,h} + \left(\frac{a_{j+1,h} - a_{j,h}}{h}\right)(x - jh),$$

$$\text{for } x \in [jh, (j+1)h), \ j \in \{0,\ldots,N\}, \tag{3.6}$$

$$\mathbb{Q}_h(a_h)(x) = \begin{cases} a_{j,h} \text{ for } x \in [(j-1/2)h, (j+1/2)h), \ j \in \{1,\ldots,N\}, \\ 0 \text{ for } x \in [0,h/2) \cup [(N+1/2)h, 1], \end{cases} \quad (3.7)$$

with the conventions $a_{0,h} = a_{N+1,h} = 0$.

The range of the extension operator $\mathbb{P}_h$ is the set of continuous, piecewise affine functions with (C^1) singularities in the points jh and vanishing on the boundary. This corresponds to the most natural approximation leading to $H_0^1(0,1)$ functions and to the point of view of the $P1$ finite element method. By the contrary, $\mathbb{Q}_h$ provides the simplest piecewise constant extension of the discrete function which, obviously, lies in $L^2(0,1)$ but not in $H_0^1(0,1)$.

Note that the extensions $\mathbb{F}_h(a_h)$ obtained using the Fourier representation (3.5) and $\mathbb{P}_h(a_h)$ do not coincide. However, they are closely related as follows:

Proposition 3.1. *For each $h = 1/(N+1) > 0$, let a_h be a sequence of discrete functions.*

Then, for $s \in \{0,1\}$, the sequence of Fourier extensions $(\mathbb{F}_h(a_h))_{h>0}$ converges strongly (respectively weakly) in $H^s(0,1)$ if and only if the sequence $(\mathbb{P}_h(a_h))_{h>0}$ converges strongly (respectively weakly) in $H^s(0,1)$. Besides, if one of these sequences converge, then they have the same limit.

Moreover, there exists a constant C independent of $h > 0$ such that

$$\frac{1}{C}\|\mathbb{F}_h(a_h)\|_{L^2} \leq \|\mathbb{P}_h(a_h)\|_{L^2} \leq C\|\mathbb{F}_h(a_h)\|_{L^2}, \quad (3.8)$$

$$\frac{1}{C}\|\mathbb{F}_h(a_h)\|_{H_0^1} \leq \|\mathbb{P}_h(a_h)\|_{H_0^1} \leq C\|\mathbb{F}_h(a_h)\|_{H_0^1}. \quad (3.9)$$

Proof. Let us begin with the case $s = 0$.

Let us first compare the $L^2(0,1)$-norms of the functions $\mathbb{F}_h(a_h)$ and $\mathbb{P}_h(a_h)$.

From the orthogonality properties of w^k (see Lemma 2.1), we have

$$\|\mathbb{F}_h(a_h)\|_{L^2(0,1)}^2 = \sum_{k=1}^N |\hat{a}_{k,h}|^2 = h\sum_{j=1}^N |a_{j,h}|^2. \quad (3.10)$$

Computing the $L^2(0,1)$-norm of $\mathbb{P}_h(a_h)$ is slightly more technical:

$$\int_0^1 |\mathbb{P}_h(a_h)(x)|^2 \, dx = \sum_{j=0}^N \int_0^h \left|a_{j,h} + x\left(\frac{a_{j+1,h} - a_{j,h}}{h}\right)\right|^2 dx$$

$$= h\sum_{j=0}^N \left[a_{j,h}^2 + a_{j,h}(a_{j+1,h} - a_{j,h}) + \frac{1}{3}(a_{j+1,h} - a_{j,h})^2\right]$$

$$= \frac{h}{3}\sum_{j=0}^N (a_{j,h}^2 + a_{j+1,h}^2 + a_{j,h}a_{j+1,h})$$

$$= \frac{h}{6}\sum_{j=0}^N (a_{j,h}^2 + a_{j+1,h}^2 + 2a_{j,h}a_{j+1,h}) + \frac{h}{6}\sum_{j=0}^N (a_{j,h}^2 + a_{j+1,h}^2)$$

$$= \frac{h}{6}\sum_{j=0}^N (a_{j,h} + a_{j+1,h})^2 + \frac{h}{3}\sum_{j=1}^N |a_{j,h}|^2. \quad (3.11)$$

It follows that the $L^2(0,1)$-norms of $\mathbb{F}_h(a_h)$ and $\mathbb{P}_h(a_h)$ are equivalent, hence implying Eq. (3.8), and then the boundedness properties for these sequences are equivalent.

This also implies that the sequence $(\mathbb{F}_h(a_h))_{h>0}$ is a Cauchy sequence in $L^2(0,1)$ if and only if the sequence $(\mathbb{P}_h(a_h))$ is a Cauchy sequence in $L^2(0,1)$, and then one of these sequences converges strongly if and only if the other one does.

To guarantee that these sequences have the same limit when they converge, we have to check that their difference, if uniformly bounded, weakly converges to zero when $h \to 0$.

Let ψ denote a smooth test function. On one hand, we have

$$\int_0^1 \mathbb{F}_h(a_h)(x)\,\psi(x)\,dx = \sum_{k=1}^N \hat{a}_{k,h} \int_0^1 w^k(x)\,\psi(x)\,dx.$$

On the other one, we have

$$\int_0^1 \mathbb{P}_h(a_h)(x)\,\psi(x)\,dx = \sum_{j=1}^N \int_{jh}^{(j+1)h} \left(a_{j,h} + \frac{a_{j+1,h} - a_{j,h}}{h}(x - jh) \right) \psi(x)\,dx$$

$$= h \sum_{j=1}^N a_{j,h}\tilde{\psi}_{j,h},$$

with

$$\tilde{\psi}_{j,h} = \frac{1}{h}\int_{(j-1)h}^{jh} \psi(x)\left(\frac{x - (j-1)h}{h} \right) dx + \frac{1}{h}\int_{jh}^{(j+1)h} \psi(x)\left(1 - \frac{x - jh}{h} \right) dx$$

$$= \frac{1}{h}\int_{(j-1)h}^{(j+1)h} \psi(x)\left(1 - \frac{|x - jh|}{h} \right) dx.$$

Using Eq. (3.4), we obtain

$$\int_0^1 \mathbb{P}_h(a_h)(x)\,\psi(x)\,dx = \sum_{k=1}^N \hat{a}_{k,h}\left(h \sum_{j=1}^N w_j^k\,\tilde{\psi}_{j,h} \right). \tag{3.12}$$

Therefore,

$$\int_0^1 (\mathbb{P}_h(a_h)(x) - \mathbb{F}_h(a_h)(x))\,\psi(x)\,dx$$

$$= \sum_{k=1}^N \hat{a}_{k,h}\left(h \sum_{j=1}^N w_j^k\,\tilde{\psi}_{j,h} - \int_0^1 w^k(x)\,\psi(x)\,dx \right). \tag{3.13}$$

Now, fix $\ell \in \mathbb{N}$, and choose $\psi(x) = w^\ell(x) = \sqrt{2}\sin(\ell \pi x)$. In this case, using Taylor's formula, we easily check that

$$\sup_{j\in\{1,\dots,N\}} |\tilde{\psi}_{j,h} - \psi(jh)| \leq \ell h\pi.$$

Since, for $\ell \leq N$, see Lemma 2.1,

$$\int_0^1 w^k(x)\,w^\ell(x)\,dx = h\sum_{j=1}^N w_j^k w^\ell(jh) = \delta_k^\ell,$$

we then obtain from Eq. (3.13) that for all $\ell \in \mathbb{N}$,

$$\int_0^1 (\mathbb{P}_h(a_h)(x) - \mathbb{F}_h(a_h)(x))\, w^\ell(x)\,dx \underset{h\to 0}{\longrightarrow} 0.$$

Since the set $\{w^\ell\}_{l\in\mathbb{N}}$ spans the whole space $L^2(0,1)$, if one of the sequences $(\mathbb{F}_h(a_h))$ or $(\mathbb{P}_h(a_h))$ converges weakly in $L^2(0,1)$, then the other one also converges weakly in $L^2(0,1)$ and has the same limit.

This completes the proof in the case $s = 0$.

We now deal with the case $s = 1$. First remark that

$$\int_0^1 |\partial_x \mathbb{F}_h(a_h)|^2 dx = \sum_{k=1}^N |\hat{a}_{k,h}|^2 k^2\pi^2 \tag{3.14}$$

from the Fourier orthogonality properties, and, using Lemma 2.1,

$$\int_0^1 |\partial_x \mathbb{P}_h(a_h)(x)|^2 dx = h\sum_{j=0}^N \left(\frac{a_{j+1,h} - a_{j,h}}{h}\right)^2 = \sum_{k=1}^N \lambda_k(h)|\hat{a}_{k,h}|^2. \tag{3.15}$$

Since $c_1 k^2 \leq \lambda_k(h) \leq c_2 k^2$, these two norms are equivalent, hence implying Eq. (3.9), and therefore the $H_0^1(0,1)$-boundedness properties of the sequences $(\mathbb{F}_h(a_h))$ and $(\mathbb{P}_h(a_h))$ are equivalent.

If one of these sequences weakly converges in $H_0^1(0,1)$, then the other one is bounded in $H_0^1(0,1)$ and weakly converges in $L^2(0,1)$ to the same limit from the previous result and then also weakly converges in $H_0^1(0,1)$.

Besides, if one of these sequences strongly converges in $H_0^1(0,1)$, it is a Cauchy sequence in $H_0^1(0,1)$, and then the other one also is a Cauchy sequence in $H_0^1(0,1)$ and therefore also strongly converges.   □

Similarly, one can prove the following:

Proposition 3.2. *For each $h = 1/(N+1) > 0$, let a_h be a sequence of discrete functions.*

Then the sequence of Fourier extensions $(\mathbb{F}_h(a_h))_{h>0}$ converges strongly (respectively weakly) in $L^2(0,1)$ if and only if the sequence $(\mathbb{Q}_h(a_h))_{h>0}$ converges strongly (respectively weakly) in $L^2(0,1)$. Besides, when they converge, they have the same limit.

Moreover, there exists a constant C independent of h > 0 such that

$$\frac{1}{C}\|\mathbb{F}_h(a_h)\|_{L^2} \leq \|\mathbb{Q}_h(a_h)\|_{L^2} \leq C\|\mathbb{F}_h(a_h)\|_{L^2}. \tag{3.16}$$

The proof is very similar to the previous one and is left to the reader.

The above propositions show that the Fourier extension plays the same role as the classical extensions by continuous piecewise affine functions or by piecewise constant functions when considering convergence issues. We make the choice of considering this Fourier extension, rather than the usual ones, since it has the advantage of being smooth.

The following result is also relevant:

Proposition 3.3. *There exists a constant C independent of h > 0 such that for all solutions φ_h of Eq. (3.1):*

$$\frac{1}{C}\|(\mathbb{F}_h(\varphi_h),\mathbb{F}_h(\partial_t\varphi_h)\|_{H_0^1\times L^2} \leq E_h[\varphi_h] \leq C\|(\mathbb{F}_h(\varphi_h),\mathbb{F}_h(\partial_t\varphi_h)\|_{H_0^1\times L^2} \tag{3.17}$$

Proof. The discrete energy of a solution φ_h of Eq. (3.1) at time t exactly coincides with the $H_0^1(0,1) \times L^2(0,1)$-norm of $(\mathbb{P}_h(\varphi_h),\mathbb{Q}_h(\partial_t\varphi_h))$ at time t. Using the equivalences (3.9) and (3.16), we immediately obtain Eq. (3.17). $\square$

In the following, we will often omit the operator $\mathbb{F}_h$ from explicit notations and directly identify the discrete function $a_h = (a_{j,h})_{j\in\{1,...,N\}}$ with its continuous Fourier extension $\mathbb{F}_h(a_h)$.

3.3 Orders of Convergence for Smooth Initial Data

In this section, we consider a solution φ of Eq. (3.2) with initial data $(\varphi^0,\varphi^1) \in H^2 \cap H_0^1(0,1) \times H_0^1(0,1)$. The solution φ of Eq. (3.2) then belongs to the space

$$\varphi \in C([0,T];H^2 \cap H_0^1(0,1)) \cap C^1([0,T];H_0^1(0,1)) \cap C^2([0,T];L^2(0,1)).$$

In order to prove it, one can remark that the energy

$$E[\varphi](t) = \int_0^1 \left(|\partial_t\varphi(t,x)|^2 + |\partial_x\varphi(t,x)|^2\right) dx$$

is constant in time for solutions of Eq. (3.2) with initial data in $H_0^1(0,1) \times L^2(0,1)$. We then apply it to $\partial_t\varphi$, which is a solution of Eq. (3.2) with initial data $(\varphi_1,\partial_{xx}\varphi_0) \in H_0^1(0,1) \times L^2(0,1)$.

The goal of this section is to prove the following result:

Proposition 3.4. *Let $(\varphi^0,\varphi^1) \in H^2 \cap H_0^1(0,1) \times H_0^1(0,1)$. Then there exist a constant $C = C(T)$ independent of (φ^0,φ^1) and a sequence $(\varphi_h^0,\varphi_h^1)$ of discrete initial data such that for all $h > 0$,*

$$\left\|(\varphi_h^0, \varphi_h^1) - (\varphi^0, \varphi^1)\right\|_{H_0^1 \times L^2} \leq Ch^{2/3} \left\|(\varphi^0, \varphi^1)\right\|_{H^2 \cap H_0^1 \times H_0^1} \qquad (3.18)$$

and the solutions φ of Eq. (3.2) with initial data (φ^0, φ^1) and φ_h of Eq. (3.1) with initial data $(\varphi_h^0, \varphi_h^1)$ satisfy, for all $h > 0$ and $t \in [0, T]$,

$$\left\|(\varphi_h(t), \partial_t \varphi_h(t)) - (\varphi(t), \partial_t \varphi(t))\right\|_{H_0^1 \times L^2} \leq Ch^{2/3} \left\|(\varphi^0, \varphi^1)\right\|_{H^2 \cap H_0^1 \times H_0^1}, \qquad (3.19)$$

and

$$\left\|\frac{\varphi_{N,h}(\cdot)}{h} + \partial_x \varphi(\cdot, 1)\right\|_{L^2(0,T)} \leq Ch^{2/3} \left\|(\varphi^0, \varphi^1)\right\|_{H^2 \cap H_0^1 \times H_0^1}. \qquad (3.20)$$

Remark 3.1. The result in Eq. (3.18) may appear somewhat surprising since when approximating $(\varphi^0, \varphi^1) \in H^2 \cap H_0^1(0,1) \times H_0^1(0,1)$ by the classical continuous piecewise affine approximations or truncated Fourier series, the approximations $(\varphi_h^0, \varphi_h^1)$ satisfy

$$\left\|(\varphi_h^0, \varphi_h^1) - (\varphi^0, \varphi^1)\right\|_{H_0^1 \times L^2} \leq Ch \left\|(\varphi^0, \varphi^1)\right\|_{H^2 \cap H_0^1 \times H_0^1} \qquad (3.21)$$

instead of Eq. (3.18).

However, the result in [45] indicates that, even if the convergence of the initial data is as in Eq. (3.21), one cannot obtain a better result than Eq. (3.19). This is due to the distance between the continuous and space semi-discrete semigroups generated by Eqs. (3.2) and (3.1), respectively, and their purely conservative nature. To be more precise, when looking at the dispersion diagram, the eigenvalues of the semi-discrete wave equation (3.1) are of the form

$$\sqrt{\lambda_k(h)} = \frac{2}{h} \sin\left(\frac{k\pi h}{2}\right),$$

whereas the ones of the continuous equation (3.2) are $\sqrt{\lambda_k} = k\pi$. In particular, for any $\varepsilon > 0$,

$$\sup_{k \leq h^{-2/3+\varepsilon}} \left\{\left|\sqrt{\lambda_k(h)} - k\pi\right|\right\} = 0, \quad \text{while} \quad \sup_{k \geq h^{-2/3-\varepsilon}} \left\{\left|\sqrt{\lambda_k(h)} - k\pi\right|\right\} = \infty.$$

Remark 3.2. The main issue in Proposition 3.4 is the estimate (3.20). Estimates (3.19) are rather classical in the context of finite element methods; see, e.g., [2] and the references therein.

Proof. Let $(\varphi^0, \varphi^1) \in H^2 \cap H_0^1(0,1) \times H_0^1(0,1)$. Expanding these initial data on the Fourier basis (recall that $w^k(x) = \sqrt{2}\sin(k\pi x)$), we have

$$\varphi^0 = \sum_{k=1}^{\infty} \hat{a}_k w^k, \quad \varphi^1 = \sum_{k=1}^{\infty} \hat{b}_k w^k.$$

The solution φ of Eq. (3.2) can then be computed explicitly in Fourier:

$$\varphi(t,x) = \sum_{|k|=1}^{\infty} \hat{\varphi}_k \exp(i\mu_k t) w^{|k|}, \quad \mu_k = k\pi, \quad \hat{\varphi}_k = \frac{1}{2}\left(\hat{a}_{|k|} + \frac{i\hat{b}_{|k|}}{\mu_k}\right).$$

And the condition $(\varphi^0, \varphi^1) \in H^2 \cap H_0^1(0,1) \times H_0^1(0,1)$ can be written as

$$\sum_{k=1}^{\infty}\left(k^4|\hat{\varphi}_k^0|^2 + k^2|\hat{\varphi}_k^1|^2\right) < \infty \quad \text{or, equivalently, } \sum_{|k|=1}^{\infty} k^4|\hat{\varphi}_k|^2 < \infty, \qquad (3.22)$$

and both these quantities are equivalent to the $H^2 \cap H_0^1(0,1) \times H_0^1(0,1)$-norm of the initial data (φ^0, φ^1).

We now look for a solution φ_h of Eq. (3.1) on the Fourier basis. Using that the functions w^k correspond to eigensolutions of the discrete Laplace operator for $k \le N$, one easily checks that any solution of Eq. (3.1) can be written as $\sum_{|k|=1}^{N} a_k w^{|k|} \exp(i\mu_k(h)t)$ with $\mu_k(h) = 2\sin(k\pi h/2)/h$. Keeping this in mind, we take

$$\varphi_h(t) = \sum_{|k|=1}^{n(h)} \hat{\varphi}_k \exp(i\mu_k(h)t) w^{|k|}, \qquad (3.23)$$

where $n(h)$ is an integer smaller than N that will be fixed later on.

We now compute how this solution approximates φ:

$$\|\varphi_h(t) - \varphi(t)\|_{H_0^1}^2$$

$$= \sum_{|k|=n(h)+1}^{\infty} k^2\pi^2|\hat{\varphi}_k|^2 + \sum_{|k|=1}^{n(h)} k^2\pi^2|\hat{\varphi}_k|^2 4\sin^2\left(\frac{(\mu_k(h)-\mu_k)t}{2}\right)$$

$$\le \frac{C}{n(h)^2}\sum_{|k|=n(h)+1}^{\infty} k^4\pi^4|\hat{\varphi}_k|^2 + C\sum_{|k|=1}^{n(h)}(k^4h^4)k^4\pi^4|\hat{\varphi}_k|^2$$

$$\le C\left(n(h)^4h^4 + \frac{1}{n(h)^2}\right)\|(\varphi^0,\varphi^1)\|_{H^2\cap H_0^1\times H_0^1}^2, \qquad (3.24)$$

where we have used that for some constant C independent of $h > 0$ and $k \in \{1,\dots,N\}$,

$$|\mu_k(h) - \mu_k| = \left|\frac{2}{h}\sin\left(\frac{k\pi h}{2}\right) - k\pi\right| \le Ck^3h^2,$$

and

$$\left|\sin\left(\frac{(\mu_k(h)-\mu_k)t}{2}\right)\right| \le CT|\mu_k(h) - \mu_k|.$$

The same can be done for $\partial_t \varphi_h$:

$$\|\partial_t \varphi_h(t) - \partial_t \varphi(t)\|_{L^2}^2$$

$$= \sum_{|k|=n(h)+1}^{\infty} k^2 \pi^2 |\hat{\varphi}_k|^2 + \sum_{|k|=1}^{n(h)} |\hat{\varphi}_k|^2 \left| \mu_k(h) e^{i\mu_k(h)t} - \mu_k e^{i\mu_k t} \right|^2$$

$$\leq C \left(n(h)^4 h^4 + \frac{1}{n(h)^2} \right) \|(\varphi^0, \varphi^1)\|_{H^2 \cap H_0^1 \times H_0^1}^2, \qquad (3.25)$$

where we used that

$$\left| \mu_k(h) e^{i\mu_k(h)t} - \mu_k e^{i\mu_k t} \right| \leq \left| 2k\pi \sin\left(\frac{(\mu_k(h) - \mu_k)t}{2} \right) \right| + |\mu_k(h) - \mu_k| \leq Ck^4 h^2.$$

Estimates (3.24) and (3.25) then imply Eqs. (3.18) and (3.19) when choosing $n(h) \simeq h^{-2/3}$, a choice that, as we will see below, also optimizes the convergence of the normal derivatives.

We shall now prove Eq. (3.20). This will be done in two main steps, computing separately the integrals

$$I_1 = \int_0^T \left| \partial_x \varphi_h(t, 1) + \frac{\varphi_{N,h}(t)}{h} \right|^2 dt, \quad \text{and} \quad I_2 = \int_0^T |\partial_x \varphi(t, 1) - \partial_x \varphi_h(t, 1)|^2 dt. \qquad (3.26)$$

Estimates on I_1. We shall first write the admissibility inequality proved in Theorem 2.1 in terms of Fourier series.

Consider a solution ϕ_h of Eq. (3.1) and write it as

$$\phi_h(t) = \sum_{|k|=1}^{N} \hat{\phi}_{k,h} e^{i\mu_k(h)t} w^{|k|},$$

where

$$\hat{\phi}_{k,h} = \frac{1}{2} \left(\hat{\phi}_{k,h}^0 + \frac{\hat{\phi}_{k,h}^1}{i\mu_k(h)} \right).$$

The energy of the solution is then given by

$$E_h = 2 \sum_{|k|=1}^{N} \lambda_{|k|}(h) \left| \hat{\phi}_{k,h} \right|^2.$$

Hence the admissibility result in Theorem 2.1 reads as follows: for any sequence $(\hat{\phi}_{k,h})$,

$$\int_0^T \left| \sum_{|k|=1}^{N} \hat{\phi}_{k,h} e^{i\mu_k(h)t} \frac{w_N^{|k|}}{h} \right|^2 dt \leq C \sum_{|k|=1}^{N} \lambda_{|k|}(h) \left| \hat{\phi}_{k,h} \right|^2. \qquad (3.27)$$

But the difference $\partial_x \varphi_h(t, 1) + \varphi_{N,h}/h$ reads as

$$
\partial_x \varphi_h(t, 1) + \frac{\varphi_{N,h}}{h}(t) = \sum_{|k|=1}^{n(h)} \hat{\varphi}_k e^{i\mu_k(h)t} \left(\partial_x w^{|k|}(1) + \frac{w_N^{|k|}}{h} \right)
$$

$$
= \sum_{|k|=1}^{n(h)} \hat{\varphi}_k \left(1 + \frac{h \partial_x w^{|k|}(1)}{w_N^{|k|}} \right) e^{i\mu_k(h)t} \frac{w_N^{|k|}}{h}.
$$

Thus, applying Eq. (3.27), we get

$$
\int_0^T \left| \partial_x \varphi_h(t, 1) + \frac{\varphi_{N,h}(t)}{h} \right|^2 dt \leq C \sum_{|k|=1}^{n(h)} \lambda_{|k|}(h) |\hat{\varphi}_k|^2 \left(1 + \frac{h \partial_x w^{|k|}(1)}{w_N^{|k|}} \right)^2. \quad (3.28)
$$

But for all $k \in \{1, \ldots, N\}$,

$$
\frac{h \partial_x w^k(1)}{w_N^k} = -\frac{k\pi h \cos(k\pi)}{\sin(k\pi h)\cos(k\pi)} = -\frac{k\pi h}{\sin(k\pi h)},
$$

and we thus have, for some explicit constant C independent of h and k, that for all $h > 0$ and $k \in \{1, \ldots, N\}$,

$$
\left| 1 + \frac{h \partial_x w^k(1)}{w_N^k} \right| \leq C(k\pi h)^2.
$$

Plugging this last estimate into Eq. (3.28) and using $\lambda_k(h) \leq Ck^2$, we obtain

$$
I_1 = \int_0^T \left| \partial_x \varphi_h(t, 1) + \frac{\varphi_{N,h}(t)}{h} \right|^2 dt \leq C \sum_{|k|=1}^{n(h)} |\hat{\varphi}_k|^2 k^6 h^4
$$

$$
\leq Cn(h)^2 h^4 \sum_{|k|=1}^{n(h)} k^4 |\hat{\varphi}_k|^2
$$

$$
\leq Cn(h)^2 h^4 \left\| (\varphi^0, \varphi^1) \right\|^2_{H^2 \cap H_0^1 \times H_0^1}. \quad (3.29)
$$

Estimates on I_2. The idea now is to see φ_h as a solution of Eq. (3.1) up to a perturbation. Note that this is a classical technique in numerical analysis and more particularly in *a posteriori* error analysis.

Indeed, recall that

$$
\varphi_h = \sum_{|k|=1}^{n(h)} \hat{\varphi}_k e^{i\mu_k(h)t} w^{|k|}(x).
$$

This implies that

$$
\partial_{tt} \varphi_h - \partial_{xx} \varphi_h = f_h, \quad (t, x) \in \mathbb{R} \times (0, 1)
$$

with

$$f_h(x,t) = \sum_{|k|=1}^{n(h)} \hat{\varphi}_k e^{i\mu_k(h)t} w^{|k|}(x) \left(-\lambda_{|k|}(h) + k^2\pi^2 \right).$$

In particular, for all $t \in \mathbb{R}$,

$$\|f_h(t)\|_{L^2(0,1)}^2 \leq \sum_{|k|=1}^{n(h)} k^4\pi^4 |\hat{\varphi}_k|^2 \left(1 - \frac{4}{k^2\pi^2 h^2} \sin^2\left(\frac{k\pi h}{2} \right) \right)^2$$

$$\leq C \sum_{|k|=1}^{n(h)} k^4\pi^4 |\hat{\varphi}_k|^2 (k\pi h)^4$$

$$\leq Cn(h)^4 h^4 \sum_{|k|=1}^{n(h)} k^4\pi^4 |\hat{\varphi}_k|^2$$

$$\leq Cn(h)^4 h^4 \left\| (\varphi^0, \varphi^1) \right\|_{H^2 \cap H_0^1 \times H_0^1}^2,$$

where the constant C is independent of $h > 0$.

Now, consider $z_h = \varphi_h - \varphi$. Then z_h satisfies the following system of equations:

$$\begin{cases} \partial_{tt}z_h - \partial_{xx}z_h = f_h, & t \in \mathbb{R}, x \in (0,1) \\ z_h(t,0) = z_h(t,1) = 0, & t \in \mathbb{R}, \\ z_h(0,x) = z_h^0(x), \; \partial_t z_h(0,x) = z^1(x), \; 0 < x < 1, \end{cases} \quad (3.30)$$

with $(z_h^0, z_h^1) = (\varphi_h^0, \varphi_h^1) - (\varphi^0, \varphi^1)$, which satisfies, according to Eqs. (3.24) and (3.25) for $t = 0$,

$$\left\| (z_h^0, z_h^1) \right\|_{H_0^1 \times L^2}^2 \leq C \left(\frac{1}{n(h)^2} + n(h)^4 h^4 \right) \left\| (\varphi^0, \varphi^1) \right\|_{H^2 \cap H_0^1 \times H_0^1}^2.$$

But this is now the continuous wave equation and one can easily check that the normal derivative of z_h then satisfies the following admissibility result: for some constant C independent of $h > 0$,

$$\int_0^T |\partial_x z_h(t,1)|^2 \, dt \leq C \left(\|f_h\|_{L^1(0,T;L^2(0,1))}^2 + \left\| (z_h^0, z_h^1) \right\|_{H_0^1 \times L^2}^2 \right).$$

For a proof of that fact we refer to the book of Lions [36] and the article [34].

This gives

$$I_2 = \int_0^T |\partial_x \varphi(t,1) - \partial_x \varphi_h(t,1)|^2 \, dt$$

$$\leq C \left(\frac{1}{n(h)^2} + n(h)^4 h^4 \right) \left\| (\varphi^0, \varphi^1) \right\|_{H^2 \cap H_0^1 \times H_0^1}^2. \quad (3.31)$$

Combining the estimates (3.29) and (3.31), we obtain

$$\int_0^T \left| \partial_x \varphi(t,1) + \frac{\varphi_{N,h}(t)}{h} \right|^2 dt \leq C \left(\frac{1}{n(h)^2} + n(h)^4 h^4 \right) \left\| (\varphi^0, \varphi^1) \right\|_{H^2 \cap H_0^1 \times H_0^1}^2 .$$

The choice $n(h) \simeq h^{-2/3}$ optimizes this estimate and yields Eq. (3.20). This choice also optimizes estimates (3.24) and (3.25) and implies Eqs. (3.18) and (3.19) and thus completes the proof.   □

3.4 Further Convergence Results

3.4.1 Strongly Convergent Initial Data

As a corollary to Proposition 3.4, we can give convergence results for *any* sequence of discrete initial data $(\varphi_h^0, \varphi_h^1)$ satisfying

$$\lim_{h \to 0} \left\| (\varphi_h^0, \varphi_h^1) - (\varphi^0, \varphi^1) \right\|_{H_0^1 \times L^2} = 0. \tag{3.32}$$

Proposition 3.5. *Let $(\varphi^0, \varphi^1) \in H_0^1(0,1) \times L^2(0,1)$ and consider a sequence of discrete initial data $(\varphi_h^0, \varphi_h^1)$ satisfying Eq. (3.32). Then the solutions φ_h of Eq. (3.1) with initial data $(\varphi_h^0, \varphi_h^1)$ converge strongly in $C([0,T]; H_0^1(0,1)) \cap C^1([0,T]; L^2(0,1))$ towards the solution φ of Eq. (3.2) with initial data (φ^0, φ^1) as $h \to 0$. Moreover, we have*

$$\lim_{h \to 0} \int_0^T \left| \partial_x \varphi(t,1) + \frac{\varphi_{N,h}}{h} \right|^2 dt = 0. \tag{3.33}$$

Proof. Let $(\varphi^0, \varphi^1) \in H_0^1(0,1) \times L^2(0,1)$ and, given $\varepsilon > 0$, choose $(\psi^0, \psi^1) \in H^2 \cap H_0^1(0,1) \times H_0^1(0,1)$ so that

$$\left\| (\varphi^0, \varphi^1) - (\psi^0, \psi^1) \right\|_{H_0^1 \times L^2} \leq \varepsilon.$$

We now use the discrete initial data (ψ_h^0, ψ_h^1) provided by Proposition 3.4. The solutions ψ_h of Eq. (3.1) with initial data (ψ_h^0, ψ_h^1) thus converge to the solution ψ of Eq. (3.2) with initial data (ψ^0, ψ^1) in the sense of Eqs. (3.19)–(3.20).

We now denote by φ_h the solutions of Eq. (3.1) with initial data $(\varphi_h^0, \varphi_h^1)$ and φ the solution of Eq. (3.2) with initial data (φ^0, φ^1).

Since $\varphi_h - \psi_h$ is a solution of Eq. (3.1), the conservation of the energy and the uniform admissibility property (2.12) yield

$$\sup_{t\in[0,T]} \|(\varphi_h, \partial_t \varphi_h)(t) - (\psi_h, \partial_t \psi_h)(t)\|_{H_0^1 \times L^2} + \left\|\frac{\varphi_{N,h} - \psi_{N,h}}{h}\right\|_{L^2(0,T)}$$

$$\leq C \|(\varphi_h^0, \varphi_h^1) - (\psi_h^0, \psi_h^1)\|_{H_0^1 \times L^2}$$

$$\leq C(\|(\varphi_h^0, \varphi_h^1) - (\varphi^0, \varphi^1)\|_{H_0^1 \times L^2} + \|(\varphi^0, \varphi^1) - (\psi^0, \psi^1)\|_{H_0^1 \times L^2}$$

$$+ \|(\psi^0, \psi^1) - (\psi_h^0, \psi_h^1)\|_{H_0^1 \times L^2})$$

$$\leq C(\|(\varphi_h^0, \varphi_h^1) - (\varphi^0, \varphi^1)\|_{H_0^1 \times L^2} + \varepsilon + C_\varepsilon h^{2/3} \|(\psi^0, \psi^1)\|_{H^2 \cap H_0^1 \times H_0^1}).$$

Besides, recalling that ψ_h converge to ψ in the sense of Eqs. (3.19)–(3.20), we have

$$\lim_{h\to 0} \sup_{t\in[0,T]} \|(\psi_h, \partial_t \psi_h)(t) - (\psi, \partial_t \psi)(t)\|_{H_0^1 \times L^2} + \left\|\partial_x \psi(t,1) + \frac{\psi_{N,h}}{h}\right\|_{L^2(0,T)} = 0.$$

We also use that the energy of the continuous wave equation (3.2) is constant in time and the admissibility result of the continuous wave equation and apply it to $\varphi - \psi$:

$$\sup_{t\in[0,T]} \|(\varphi, \partial_t \varphi)(t) - (\psi, \partial_t \psi)(t)\|_{H_0^1 \times L^2} + \|\partial_x \varphi(t,1) - \partial_x \psi(t,1)\|_{L^2(0,T)} \leq C\varepsilon.$$

Combining these three estimates and taking the limsup as $h \to 0$, for all $\varepsilon > 0$, we get

$$\limsup_{h\to 0} \left(\sup_{t\in[0,T]} \|(\varphi_h, \partial_t \varphi_h)(t) - (\varphi, \partial_t \varphi)(t)\|_{H_0^1 \times L^2} \right.$$

$$\left. + \left\|\frac{\varphi_{N,h}(t)}{h} + \partial_x \varphi(t,1)\right\|_{L^2(0,T)} \right) \leq C\varepsilon.$$

This concludes the proof of Proposition 3.5 since $\varepsilon > 0$ was arbitrary.   □

3.4.2 Smooth Initial Data

In this section, we derive higher convergence rates when the initial data are smoother. In order to do that, we introduce, for $\ell \in \mathbb{R}$, the functional space $H_{(0)}^\ell$ defined by

$$H_{(0)}^\ell(0,1) = \left\{ \varphi = \sum_{k=1}^\infty \hat{\varphi}_k w^k, \text{ with } \sum_{k=1}^\infty k^{2\ell} |\hat{\varphi}_k|^2 < \infty \right\}$$

endowed with the norm $\|\varphi\|_{H_{(0)}^\ell}^2 = \sum_{k=1}^\infty k^{2\ell} |\hat{\varphi}_k|^2.$ (3.34)

These functional spaces correspond to the domains $\mathscr{D}((-\Delta_d)^{\ell/2})$ of the fractional powers of the Dirichlet Laplace operator $-\Delta_d$. In particular, we have $H_{(0)}^0(0,1) = L^2(0,1)$, $H_{(0)}^1(0,1) = H_0^1(0,1)$ and $H_{(0)}^{-1}(0,1) = H^{-1}(0,1)$.

As an extension of Proposition 3.4, we obtain:

Proposition 3.6. *Let $\ell \in (0,3]$ and $(\varphi^0, \varphi^1) \in H_{(0)}^{\ell+1}(0,1) \times H_{(0)}^{\ell}(0,1)$. Denote by φ the solution of Eq. (3.2) with initial data (φ^0, φ^1). Then there exists a constant $C = C(T,\ell)$ independent of (φ^0, φ^1) such that the sequence φ_h of solutions of Eq. (3.1) with initial data $(\varphi_h^0, \varphi_h^1)$ constructed in Proposition 3.4 satisfies, for all $h > 0$,*

$$\sup_{t\in[0,T]} \|(\varphi_h(t), \partial_t \varphi_h(t)) - (\varphi(t), \partial_t \varphi(t))\|_{H_0^1 \times L^2}$$

$$\leq Ch^{2\ell/3} \|(\varphi^0, \varphi^1)\|_{H_{(0)}^{\ell+1} \times H_{(0)}^{\ell}}, \tag{3.35}$$

and

$$\left\| \frac{\varphi_{N,h}(\cdot)}{h} + \partial_x \varphi(\cdot, 1) \right\|_{L^2(0,T)} \leq Ch^{2\ell/3} \|(\varphi^0, \varphi^1)\|_{H_{(0)}^{\ell+1} \times H_{(0)}^{\ell}}. \tag{3.36}$$

In particular, for $\ell = 3$, this result reads as follows: if $(\varphi^0, \varphi^1) \in H_{(0)}^4(0,1) \times H_{(0)}^3(0,1)$, the sequence φ_h constructed in Proposition 3.4 satisfies the following convergence results:

$$\sup_{t\in[0,T]} \|(\varphi_h(t), \partial_t \varphi_h(t)) - (\varphi(t), \partial_t \varphi(t))\|_{H_0^1 \times L^2} \leq Ch^2 \|(\varphi^0, \varphi^1)\|_{H_{(0)}^4 \times H_{(0)}^3}, \tag{3.37}$$

and

$$\left\| \frac{\varphi_{N,h}(\cdot)}{h} + \partial_x \varphi(\cdot, 1) \right\|_{L^2(0,T)} \leq Ch^2 \|(\varphi^0, \varphi^1)\|_{H_{(0)}^4 \times H_{(0)}^3}. \tag{3.38}$$

Note that we cannot expect to go beyond the rate h^2 since the method is consistent of order 2.

Proof (Sketch). The proof of these convergence results follows line to line the one of Proposition 3.4.

Let us for instance explain how it has to be modified to get Eq. (3.37). First remark that Eq. (3.22) now reads

$$\sum_{|k|=1}^{\infty} k^{2\ell+2} |\hat{\phi}_k|^2 \simeq \|(\varphi^0, \varphi^1)\|_{H_{(0)}^{\ell+1} \times H_{(0)}^{\ell}}^2.$$

Estimates (3.24)–(3.25) can then be modified into

$$\|\varphi_h(t) - \varphi(t)\|_{H_0^1}^2 + \|\partial_t \varphi_h(t) - \partial_t \varphi(t)\|_{L^2}^2$$

$$\leq C \left(h^4 n(h)^{6-2\ell} + \frac{1}{n(h)^{2\ell}} \right) \|(\varphi^0, \varphi^1)\|_{H_{(0)}^{\ell+1} \times H_{(0)}^{\ell}}^2,$$

thus implying Eq. (3.35) immediately when taking $n(h) \simeq h^{-2/3}$.

The proof of the strong convergence (3.36) also relies upon the estimate

$$I_1 + I_2 \leq C \left(h^4 n(h)^{6-2\ell} + \frac{1}{n(h)^{2\ell}} \right) \|(\varphi^0, \varphi^1)\|^2_{H_{(0)}^{\ell+1} \times H_{(0)}^{\ell}},$$

where I_1 and I_2 are, respectively, given as above by Eq. (3.26). Details are left to the reader. $\quad\square$

3.4.3 General Initial Data

In Propositions 3.4 and 3.6, the discrete initial data are very special ones constructed during the proof. In this section, we explain how this yields convergence rates even for other initial data.

Proposition 3.7. *Let $\ell \in (0,3]$ and $(\varphi^0, \varphi^1) \in H_{(0)}^{\ell+1}(0,1) \times H_{(0)}^{\ell}(0,1)$ and consider a sequence (ϕ_h^0, ϕ_h^1) satisfying, for some constants $C_0 > 0$ and $\theta > 0$ independent of $h > 0$,*

$$\|(\phi_h^0, \phi_h^1) - (\varphi^0, \varphi^1)\|_{H_0^1 \times L^2} \leq C_0 h^\theta. \tag{3.39}$$

Denote by ϕ_h (respectively φ) the solution of Eq. (3.1) (resp. Eq. (3.2)) with initial data (ϕ_h^0, ϕ_h^1) (resp. (φ^0, φ^1)).
Then the following estimates hold:

$$\sup_{t \in [0,T]} \|(\phi_h(t), \partial_t \phi_h(t)) - (\varphi(t), \partial_t \varphi(t))\|_{H_0^1 \times L^2}$$

$$\leq C \left(h^{2\ell/3} \|(\varphi^0, \varphi^1)\|_{H_{(0)}^{\ell+1} \times H_{(0)}^{\ell}} + C_0 h^\theta \right), \tag{3.40}$$

and

$$\left\| \frac{\phi_{N,h}(\cdot)}{h} + \varphi_x(\cdot, 1) \right\|_{L^2(0,T)} \leq C \left(h^{2\ell/3} \|(\varphi^0, \varphi^1)\|_{H_{(0)}^{\ell+1} \times H_{(0)}^{\ell}} + C_0 h^\theta \right). \tag{3.41}$$

Proof. The proof easily follows from Proposition 3.6 since it simply consists in comparing φ_h, the solution of Eq. (3.1) given by Proposition 3.4, and ϕ_h, the solution of Eq. (3.1) with initial data (ϕ_h^0, ϕ_h^1). But $\varphi_h - \phi_h$ solves Eq. (3.1) with an initial data of $H_0^1(0,1) \times L^2(0,1)$-norm less than $Ch^{2\ell/3}\|(\varphi^0, \varphi^1)\|_{H_{(0)}^{\ell+1} \times H_{(0)}^{\ell}} + CC_0 h^\theta$.

The first estimate (3.40) then follows immediately from the fact that the discrete energy is constant for solutions of Eq. (3.1), whereas estimate (3.41) is based on the uniform admissibility results proved in Theorem 2.1. $\quad\square$

3.4.4 Convergence Rates in Weaker Norms

For later use, we also give the following result:

Proposition 3.8. *Let* $(\varphi^0, \varphi^1) \in H^2_{(0)}(0,1) \times H^1_{(0)}(0,1)$. *Denote by* φ *the solution of Eq. (3.2) with initial data* (φ^0, φ^1). *Then for all* $\ell \in (0,3]$, *there exists a constant* $C = C(T, \ell)$ *independent of* (φ^0, φ^1) *such that the sequence* φ_h *of solutions of Eq. (3.1) with initial data* $(\varphi^0_h, \varphi^1_h)$ *constructed in Proposition 3.4 satisfies, for all* $h > 0$,

$$
\sup_{t \in [0,T]} \left\| (\varphi_h(t), \partial_t \varphi_h(t), \partial_{tt} \varphi_h(t)) - (\varphi(t), \partial_t \varphi(t), \partial_{tt} \varphi(t)) \right\|_{H^{2-\ell}_{(0)} \times H^{1-\ell}_{(0)} \times H^{-\ell}_{(0)}}
$$
$$
\leq Ch^{2\ell/3} \left\| (\varphi^0, \varphi^1) \right\|_{H^2_{(0)} \times H^1_{(0)}}. \tag{3.42}
$$

In particular, if (ϕ^0_h, ϕ^1_h) *are discrete functions such that for some* $\ell_0 \in (0,3]$, C_0 *independent of* $h > 0$ *and* $\theta > 0$,

$$
\left\| (\phi^0_h, \phi^1_h) - (\varphi^0, \varphi^1) \right\|_{H^{2-\ell_0}_{(0)} \times H^{1-\ell_0}_{(0)}} \leq C_0 h^\theta, \tag{3.43}
$$

then denoting by ϕ_h *the corresponding solution of Eq. (3.1), we have*

$$
\sup_{t \in [0,T]} \left\| (\phi_h(t), \partial_t \phi_h(t), \partial_{tt} \phi_h(t)) - (\varphi(t), \partial_t \varphi(t), \partial_{tt} \varphi(t)) \right\|_{H^{2-\ell_0}_{(0)} \times H^{1-\ell_0}_{(0)} \times H^{-\ell_0}_{(0)}}
$$
$$
\leq C \left(h^{2\ell_0/3} \left\| (\varphi^0, \varphi^1) \right\|_{H^2_{(0)} \times H^1_{(0)}} + C_0 h^\theta \right). \tag{3.44}
$$

Proof. The proof of Eq. (3.42) again follows the one of Proposition 3.4. This time, following Eqs. (3.24)–(3.25), we get

$$
\left\| \varphi_h(t) - \varphi(t) \right\|^2_{H^{2-\ell}_{(0)}} + \left\| \partial_t \varphi_h(t) - \partial_t \varphi(t) \right\|^2_{H^{1-\ell}_{(0)}}
$$
$$
\leq C \left(n(h)^{6-2\ell} h^4 + \frac{1}{n(h)^{2\ell}} \right) \left\| (\varphi^0, \varphi^1) \right\|^2_{H^2_{(0)} \times H^1_{(0)}}.
$$

The proof of the estimate

$$
\sup_{t \in [0,T]} \left\| \partial_{tt} \varphi_h(t) - \partial_{tt} \varphi(t) \right\|_{H^{-\ell}_{(0)}} \leq Ch^{2\ell/3} \left\| (\varphi^0, \varphi^1) \right\|_{H^2_{(0)} \times H^1_{(0)}}
$$

can be done by writing

$$
\partial_{tt} \varphi_h(t) - \partial_{tt} \varphi(t) = \sum_{|k|=1}^{n(h)} \hat{\varphi}_k w^{|k|} \left(-\mu_k(h)^2 e^{i\mu_k(h)t} + \mu_k^2 e^{i\mu_k t} \right) + \sum_{n(h)+1}^{\infty} \hat{\varphi}_k w^{|k|} \mu_k^2 e^{i\mu_k t}
$$

and by using the estimate

$$\left| -\mu_k(h)^2 e^{i\mu_k(h)t} + \mu_k^2 e^{i\mu_k t} \right| \le Ck^5 h^2.$$

The complete proof of Eq. (3.42) is left to the reader.

The proof of Eq. (3.44) for initial data satisfying Eq. (3.43) is very similar to the one of Proposition 3.7 and is based on the following facts:

- For any ψ_h solution of the discrete wave equation (3.1), for all $\ell \in \mathbb{Z}$, the $H^{2-\ell}_{(0)}(0,1) \times H^{1-\ell}_{(0)}(0,1)$-norm of $(\psi_h(t), \partial_t \psi_h(t))$ is independent of the time $t \ge 0$, as one easily checks by writing the solutions under the form

$$\psi_h(t) = \sum_{k=1}^{N} w^k \left(\hat{\psi}_k e^{i\mu_k(h)t} + \hat{\psi}_{-k} e^{-i\mu_k(h)t} \right).$$

Applying this remark to $(\psi_h, \partial_t \psi_h)$ and to $(\partial_t \psi_h, \partial_{tt} \psi_h)$ for $\psi_h = \phi_h - \varphi_h$, we get

$$\sup_{t \in [0,T]} \left\| (\phi_h(t), \partial_t \phi_h(t), \partial_{tt} \phi_h(t)) - (\varphi(t), \partial_t \varphi(t), \partial_{tt} \varphi(t)) \right\|_{H^{2-\ell_0}_{(0)} \times H^{1-\ell_0}_{(0)} \times H^{-\ell_0}_{(0)}}$$

$$\le C \left(h^{2\ell_0/3} \left\| (\varphi^0, \varphi^1) \right\|_{H^2_{(0)} \times H^1_{(0)}} \right.$$

$$\left. + \left\| (\phi_h^0, \phi_h^1, \Delta_h \phi_h^0) - (\varphi_h^0, \varphi_h^1, \Delta_h \varphi_h^0) \right\|_{H^{2-\ell_0}_{(0)} \times H^{1-\ell_0}_{(0)} \times H^{-\ell_0}_{(0)}} \right).$$

- By construction,

$$\left\| \Delta_h \phi_h^0 - \Delta_h \varphi_h^0 \right\|_{H^{-\ell_0}_{(0)}} \le C \left\| \phi_h^0 - \varphi_h^0 \right\|_{H^{2-\ell_0}_{(0)}} ;$$

hence

$$\left\| (\phi_h^0, \phi_h^1, \Delta_h \phi_h^0) - (\varphi_h^0, \varphi_h^1, \Delta_h \varphi_h^0) \right\|_{H^{2-\ell_0}_{(0)} \times H^{1-\ell_0}_{(0)} \times H^{-\ell_0}_{(0)}}$$

$$\le C \left\| (\phi_h^0, \phi_h^1) - (\varphi_h^0, \varphi_h^1) \right\|_{H^{2-\ell_0}_{(0)} \times H^{1-\ell_0}_{(0)}}.$$

- We finally conclude Eq. (3.44) by using Eq. (3.43) and the estimate (3.42) for $t = 0$. $\qquad\qquad \square$

3.5 Numerics

In this section, we briefly illustrate the above convergence results on the normal derivatives. The rate of convergence of the discrete trajectories towards the continuous ones is well known and well illustrated in the literature.

We thus choose an initial data $(\varphi^0, \varphi^1) \in H^1_0(0,1) \times L^2(0,1)$.

For $N \in \mathbb{N}$, we set $h = 1/(N+1)$ and take $(\varphi_h^0, \varphi_h^1)$ defined by $\varphi_{j,h}^0 = \varphi^0(jh)$ and $\varphi_{j,h}^1 = \int_{((j-1/2)h,(j+1/2)h)} \varphi^1(jh)$ for all $j \in \{1,\ldots,N\}$. We then compute φ_h the corresponding solution of Eq. (3.1) and the corresponding discrete derivative at $x = 1$, i.e., $-\varphi_{N,h}(t)/h$.

Note that, actually, this discrete solution should rather be denoted as $\varphi_{h,\Delta t}$ since we also discretize in time using an explicit scheme. More precisely, if $\varphi_{h,\Delta t}^k$ denotes the approximation of φ_h at time $k\Delta t$, we solve

$$\varphi_h^{k+1} = 2\varphi_h^k - \varphi_h^{k-1} - (\Delta t)^2 \Delta_h \varphi_h^k. \tag{3.45}$$

The CFL condition is chosen such that $\Delta t/h = 0.2$ so that the convergence of the scheme (in what concurs solving the boundary–initial value problem) is ensured.

Since our goal is to estimate rates of convergence, we also need a reference data. In order to do that, we expand the initial data (φ^0, φ^1) in Fourier:

$$\varphi^0 = \sum_{k=1}^{\infty} \hat{a}_k w^k, \qquad \varphi^1 = \sum_{k=1}^{\infty} \hat{b}_k w^k.$$

The corresponding solution φ of Eq. (3.2) is then explicitly given by

$$\varphi(t) = \sum_{k=1}^{\infty} \left(\hat{a}_k \cos(k\pi t) + \hat{b}_k \frac{\sin(k\pi t)}{k\pi} \right) w^k,$$

so that

$$\partial_x \varphi(t,1) = \sum_{k=1}^{\infty} \left(\hat{a}_k \cos(k\pi t) + \hat{b}_k \frac{\sin(k\pi t)}{k\pi} \right) \sqrt{2}(-1)^k k\pi. \tag{3.46}$$

Of course, we cannot compute numerically these Fourier series for the continuous solutions of Eq. (3.2) since they involve infinite sums. So we take a reference number N_{ref} large enough and replace the infinite sum in formula (3.46) by a truncated version up to N_{ref}. N_{ref} is taken to be large compared to N, the number of nodes in the space discretization involved in the computations of $\varphi_{N,h}(t)/h$. We thus approximate the normal derivative by

$$(\partial_x \varphi(t,1))_{\text{ref}} = \sum_{k=1}^{N_{\text{ref}}} \left(\hat{a}_k \cos(k\pi t) + \hat{b}_k \frac{\sin(k\pi t)}{k\pi} \right) \sqrt{2}(-1)^k k\pi.$$

In the computations below, we take $N_{\text{ref}} = 1,000$ for N varying between 200 and 400.

In Fig. 3.1 (left), we have chosen (φ^0, φ^1) as follows:

$$\varphi^0(x) = \sin(\pi x), \quad \varphi^1(x) = 0. \tag{3.47}$$

In this particular case, the continuous solution involves one single Fourier mode. So, we could have taken $N_{\text{ref}} = 1$. Figure 3.1 (left) represents the $L^2(0,T)$-norm of $(\partial_x\varphi(t,1))_{\text{ref}} + \varphi_{N,h}(t)/h$ for $T = 1$ versus N in logarithmic scales. The slope of the linear regression is -1.99, thus very close to -2, the rate predicted by Proposition 3.7.

We then test the initial data

$$\varphi^0(x) = 0, \quad \varphi^1(x) = \begin{cases} -x & \text{if } x < 1/2, \\ -x+1 & \text{if } x > 1/2, \end{cases} \tag{3.48}$$

and plot the error in Fig. 3.1 (middle). The initial data velocity only belongs to $\cap_{\varepsilon>0}H_{(0)}^{1/2-\varepsilon}(0,1)$, so the predicted rate of convergence given by Proposition 3.7 is $-(1/3)^-$. This is indeed very close to the slope -0.31 observed in Fig. 3.1 (right).

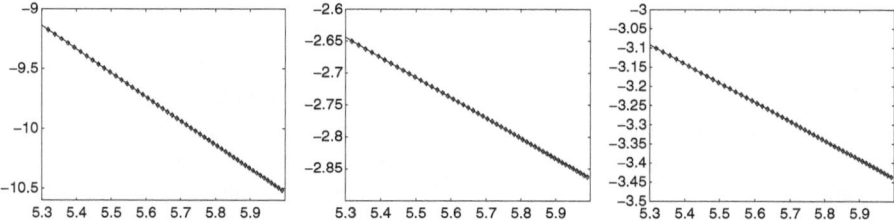

Fig. 3.1 Plot of $|(\partial_x\varphi(t,1))_{\text{ref}} + \varphi_{N,h}(t)/h|_{L^2(0,T)}$ versus $\log(N)$ for $N \in \{200,\ldots,400\}$, $N_{\text{ref}} = 1{,}000$ and $T = 1$. *Left*: for the initial data (φ^0, φ^1) in Eq. (3.47), slope of the linear regression $= -1.99$. *Middle*: for the initial data (φ^0, φ^1) in Eq. (3.48), slope $= -0.31$. *Right*: for the initial data (φ^0, φ^1) in Eq. (3.49), with $(\partial_x\varphi(t,1))_{\text{ref}} = -1+t$ in this case, slope $= -0.5$.

These numerical experiments both confirm the accuracy of the rates of convergence derived in Proposition 3.7.

We then test the initial data

$$\varphi^0(x) = 0, \quad \varphi^1(x) = x. \tag{3.49}$$

These data are smooth but $\varphi^1(1) \neq 0$. Hence φ^1 only belongs to $\cap_{\varepsilon>0}H_{(0)}^{1/2-\varepsilon}(0,1)$ and we thus expect a convergence rate of order $h^{1/3}$. Note that in this case, the normal derivative of the solution at $x = 1$ can be computed explicitly using Fourier series and $\partial_x\varphi(t,1) = -1+t$ (recall the formula (3.46)). Of course, we are thus going to use this explicit expression to compute $(\partial_x\varphi(t,1))_{\text{ref}} = -1+t$ in this case.

Note that the numerical simulations yield the slope -0.5 for the linear regression (see Fig. 3.1 (right)). This error term mainly comes from the fact that the continuous solution φ of Eq. (3.2) does not satisfy $\partial_x\varphi^0(x) = -1$ as the computation $(\partial_x\varphi(t,1))_{\text{ref}} = -1+t$ would imply for $t = 0$. This creates a layer close to $t = 0$ that the numerical method has some difficulties to handle. In Fig. 3.2, we represent the normal derivative computed numerically for $N = 300$ and compare it with the continuous normal derivative $\partial_x\varphi(t,1) = -1+t$. As one can see, there is a boundary layer close to $t = 0$.

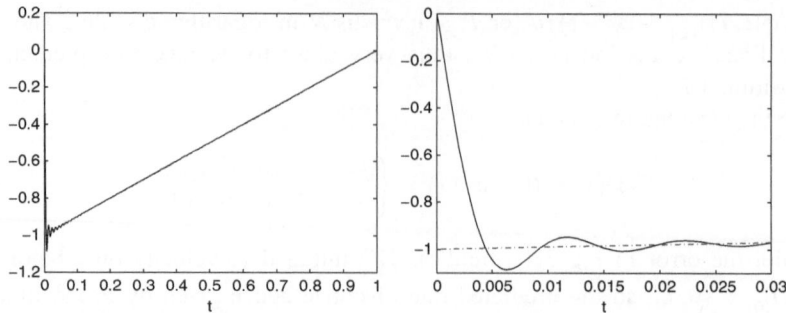

Fig. 3.2 Plot of $-\varphi_{N,h}(t)/h$ computed for $N = 300$ (*black solid line*) and of $(\partial_x\varphi(t,1))_{\mathrm{ref}} = -1+t$ (*red dash dot line*) for (φ^0, φ^1) in Eq. (3.49). *Left*: on the time interval $(0,1)$. *Right*: a zoom on the time interval $(0,0.03)$.

This last example illustrates the fact that the boundary conditions play an important role for the regularity properties of the trajectory of the continuous model (3.2) and therefore also have an influence on the rates of convergence of the corresponding approximations given by Eq. (3.1). The above example also confirms the good accuracy of the rates of convergence given in Proposition 3.7 when the regularity properties are limited by the boundary conditions.

Chapter 4
Convergence with Nonhomogeneous Boundary Conditions

4.1 The Setting

In this chapter, we consider the continuous wave equation

$$\begin{cases} \partial_{tt} y - \partial_{xx} y = 0, & (t,x) \in (0,T) \times (0,1), \\ y(t,0) = 0, \quad y(t,1) = v(t), \ t \in (0,T), \\ (y(0,\cdot), \partial_t y(0,\cdot)) = (y^0, y^1), \end{cases} \tag{4.1}$$

with

$$(y^0, y^1) \in L^2(0,1) \times H^{-1}(0,1), \qquad v \in L^2(0,T). \tag{4.2}$$

Following [36] (see also [33, 35]), system (4.1) can be solved uniquely in the sense of transposition and the solution y belongs to

$$C([0,T]; L^2(0,1)) \times C^1([0,T]; H^{-1}(0,1)).$$

Let us briefly recall the main ingredients of this definition of solution in the sense of transposition and this result.

The key idea is the following. Given smooth functions f, the solutions φ of

$$\begin{cases} \partial_{tt} \varphi - \partial_{xx} \varphi = f, & (t,x) \in (0,T) \times (0,1), \\ \varphi(t,0) = \varphi(t,1) = 0, & t \in (0,T), \\ (\varphi(T,\cdot), \partial_t \varphi(T,\cdot)) = (0,0), \end{cases} \tag{4.3}$$

which are smooth for smooth f, should satisfy

$$\int_0^T \int_0^1 y f \, dx \, dt = -\int_0^T v(t) \partial_x \varphi(t,1) \, dt$$
$$- \int_0^1 y^0(x) \partial_t \varphi(0,x) \, dx + \langle y^1, \varphi(0,\cdot) \rangle_{H^{-1}, H_0^1}. \tag{4.4}$$

S. Ervedoza and E. Zuazua, *Numerical Approximation of Exact Controls for Waves*, SpringerBriefs in Mathematics, DOI 10.1007/978-1-4614-5808-1_4, © Sylvain Ervedoza and Enrique Zuazua 2013

Thus one should first check that if $f \in L^1(0,T;L^2(0,1))$, then the solution φ of Eq. (4.3) belongs to the energy space $C([0,T];H_0^1(0,1)) \cap C^1([0,T];L^2(0,1))$ and is such that $\partial_x \varphi(t,1) \in L^2(0,T)$ with the following continuity estimate:

$$\|(\varphi, \partial_t \varphi)\|_{L^\infty(0,T;H_0^1(0,1) \times L^2(0,1))} + \|\partial_x \varphi(t,1)\|_{L^2(0,T)} \le C \|f\|_{L^1(0,T;L^2(0,1))}. \quad (4.5)$$

Of course, there, the first term can be estimated easily through the energy identity, whereas the estimate on the normal derivative of φ at $x = 1$ is a hidden regularity result that can be easily proved using multiplier techniques.

Assuming Eq. (4.5), the map

$$\mathscr{L}(f) = -\int_0^T v(t) \partial_x \varphi(t,1) \, dt - \int_0^1 y^0(x) \partial_t \varphi(0,x) \, dx + \langle y^1, \varphi(0, \cdot) \rangle_{H^{-1}, H_0^1}$$

is continuous on $L^1(0,T;L^2(0,1))$ and thus there is a unique function y in the space $L^\infty(0,T;L^2(0,1))$ that represents $\mathscr{L}$, which is by definition the solution y of Eq. (4.1) in the sense of transposition. The solution y actually belongs to the space $C([0,T];L^2(0,1))$ since it can be approximated in $L^\infty(0,T;L^2(0,1))$ by smooth functions by taking smooth approximations of v, y^0, and y^1.

A similar duality argument shows that $\partial_t y$ belongs to $C([0,T];H^{-1}(0,1))$.

Let us finally mention the following regularity result (see [34]): if $(y^0, y^1) \in H_0^1(0,1) \times L^2(0,1)$ and $v \in H^1(0,T)$ satisfies $v(0) = 0$, then the solution y of Eq. (4.1) satisfies

$$y \in C([0,T];H^1(0,1)) \cap C^1([0,T];L^2(0,1)) \text{ and } \Delta y \in C([0,T];H^{-1}(0,1)). \quad (4.6)$$

Now, the goal of this chapter is to study the convergence of the solutions of

$$\begin{cases} \partial_{tt} y_{j,h} - \dfrac{1}{h^2}(y_{j+1,h} - 2y_{j,h} + y_{j-1,h}) = 0, & (t,j) \in (0,T) \times \{1,\ldots,N\}, \\ y_{0,h} = 0, \quad y_{N+1,h}(t) = v_h(t), & t \in (0,T), \\ (y_h(0), \partial_t y_h(0)) = (y_h^0, y_h^1), \end{cases} \quad (4.7)$$

towards the solution y of Eq. (4.1), under suitable convergence assumptions on the data (y_h^0, y_h^1) and v_h to (y^0, y^1) and v.

As in Chap. 3, y_h will be identified with its Fourier extension $\mathbb{F}_h(y_h)$. This will allow us to identify the $H^{-1}(0,1)$-norm of f_h as

$$\|f_h\|_{H^{-1}(0,1)} = \|z_h\|_{H_0^1(0,1)}, \text{ where } z_h \text{ solves } -\partial_{xx} z_h = f_h \text{ on } (0,1), \quad z_h(0) = z_h(1).$$

Note that, expanding these discrete functions on the Fourier basis, one can check (see Proposition 4.1 below) that this norm is equivalent to $\|\tilde{z}_h\|_{H_0^1(0,1)}$, where $\tilde{z}_h$ solves

$$-\frac{1}{h^2}\left(\tilde{z}_{j+1,h} + \tilde{z}_{j-1,h} - 2\tilde{z}_{j,h}\right) = f_{j,h}, \quad j \in \{1,\ldots,N\}, \quad \tilde{z}_{0,h} = \tilde{z}_{N+1,h} = 0.$$

The outline of this Chap. 4 is as follows. Since we are working with the $H^{-1}(0,1)$-norm, it will be convenient to present some further convergence results for the discrete Laplace operator. In Sect. 4.3 we give some uniform bounds on the solutions y_h of Eq. (4.7). In Sect. 4.4 we derive explicit rates of convergence for smooth solutions. In Sect. 4.5 we explain how these results yield various convergence results. In Sect. 4.6, we illustrate our theoretical results by numerical experiments.

4.2 The Laplace Operator

In this section, we focus on the convergence of the discrete Laplace operator Δ_h, defined for discrete functions $z_h = (z_{j,h})_{j \in \{1,...,N\}}$ by

$$(\Delta_h z_h)_j = \frac{1}{h^2}(z_{j+1,h} - 2z_{j,h} + z_{j-1,h}), \quad j \in \{1,...,N\}, \text{ with } z_{0,h} = z_{N+1,h} = 0.$$
(4.8)

In particular, we give various results that will be used afterwards.

Let us first recall that the operator $-\Delta_h$ is self-adjoint positive definite on $\mathbb{R}^N$ according to the analysis done in Sect. 2.2. Besides, its eigenvectors w^k and eigenvalues $\lambda_k(h) = \mu_k(h)^2$ are explicit; the k-th eigenvector $w^k(x) = \sqrt{2}\sin(k\pi x)$ is independent of $h > 0$ and $\mu_k(h) = 2\sin(k\pi h/2)/h$.

4.2.1 Natural Functional Spaces

In this section, we focus on the case of "natural" functional spaces, i.e., in our case $H_0^1(0,1)$, $L^2(0,1)$, and $H^{-1}(0,1)$.

As already mentioned, we have the following:

Proposition 4.1. *If f_h is a discrete function, then there exists a constant C independent of $h \in (0,1)$ such that*

$$\frac{1}{C}\|f_h\|_{H^{-1}} \leq \left\|(-\Delta_h)^{-1}f_h\right\|_{H_0^1} \leq C\|f_h\|_{H^{-1}}.$$
(4.9)

To simplify notations, for $f \in H^{-1}(0,1)$, we shall often denote by $(-\partial_{xx})^{-1}f$ the solution $z \in H_0^1(0,1)$ of

$$-\partial_{xx}z = f \quad \text{on } (0,1), \qquad z(0) = z(1) = 0.$$

Proof. Since f_h is a discrete function, it can be expanded in Fourier series as follows:

$$f_h = \sum_{k=1}^{N} f_k w^k.$$

Then the expansions of $z = (-\partial_{xx})^{-1} f_h$ and $z_h = (-\Delta_h)^{-1} f_h$ are known:

$$z = \sum_{k=1}^{N} \frac{f_k}{\mu_k^2} w^k, \qquad z_h = \sum_{k=1}^{N} \frac{f_k}{\mu_k(h)^2} w^k.$$

Hence

$$\|z\|_{H_0^1}^2 = \sum_{k=1}^{N} \frac{|f_k|^2}{\mu_k^2}, \qquad \|z_h\|_{H_0^1}^2 = \sum_{k=1}^{N} \frac{|f_k|^2}{\mu_k^2} \frac{\mu_k^4}{\mu_k(h)^4}.$$

Since for all $k \in \{1, \ldots, N\}$,

$$1 \le \frac{\mu_k^4}{\mu_k(h)^4} \le \frac{\pi^4}{16},$$

we easily get Proposition 4.1. $\square$

We now prove the following convergence result:

Theorem 4.1. *Let $f \in L^2(0,1)$ and expand it in Fourier series as*

$$f = \sum_{k=1}^{\infty} f_k w^k, \tag{4.10}$$

and set

$$f_h = \sum_{k=1}^{N} f_k w^k. \tag{4.11}$$

Let then z be the solution of

$$-\partial_{xx} z = f, \text{ on } (0,1), \qquad z(0) = z(1) = 0, \tag{4.12}$$

and z_h of

$$-(\Delta_h z_h)_j = f_{j,h}, \quad j \in \{1, \ldots, N\}. \tag{4.13}$$

Then

$$\|f - f_h\|_{H^{-1}} + \|z - z_h\|_{H_0^1} \le Ch \|f\|_{L^2} \tag{4.14}$$

$$\|z - z_h\|_{L^2} \le Ch^2 \|f\|_{L^2}. \tag{4.15}$$

Remark 4.1. Of course, Theorem 4.1 is very classical and can be found for many different discretization schemes and in particular for finite-element methods; see for instance the textbook [46].

Proof. Our proof is of course based on the fact that the functions w^k are eigenvectors of both the continuous and discrete Laplace operators. Note that it is straightforward to check that

$$\|f - f_h\|_{H^{-1}} \le Ch \|f\|_{L^2}.$$

We thus focus on the comparison between z and z_h. Again, we use the fact that the expansions of z and z_h in Fourier are explicit:

$$z = \sum_{k=1}^{\infty} \frac{f_k}{\mu_k^2} w^k, \qquad z_h = \sum_{k=1}^{N} \frac{f_k}{\mu_k(h)^2} w^k. \tag{4.16}$$

Now, computing the H_0^1-norm of $z - z_h$ is easy:

$$\|z - z_h\|_{H_0^1}^2 = \sum_{k=1}^{N} \frac{|f_k|^2}{\mu_k^2} \left(1 - \frac{\mu_k^2}{\mu_k(h)^2}\right)^2 + \sum_{k=N+1}^{\infty} \frac{|f_k|^2}{\mu_k^2}$$

$$\leq C \sum_{k=1}^{N} |f_k|^2 k^2 h^4 + \frac{1}{N^2} \sum_{k=N+1}^{\infty} |f_k|^2,$$

where we have used that

$$\frac{1}{\mu_k^2} \left(1 - \frac{\mu_k^2}{\mu_k(h)^2}\right)^2 \leq C k^2 h^4, \quad \forall k \in \{1, \dots, N\}. \tag{4.17}$$

Hence

$$\|z - z_h\|_{H_0^1}^2 \leq C \left(N^2 h^4 + \frac{1}{N^2}\right) \|f\|_{L^2}^2.$$

Since $N + 1 = 1/h$, this concludes the proof of Eq. (4.14).

Similarly, one derives

$$\|z - z_h\|_{L^2}^2 \leq C \left(h^4 + \frac{1}{N^4}\right) \|f\|_{L^2}^2,$$

which immediately implies Eq. (4.15). $\qquad\qquad\qquad\qquad\qquad\qquad\qquad\square$

From Proposition 4.1 and Theorem 4.1 we deduce:

Theorem 4.2. *Let $f \in H^{-1}(0,1)$ and f_h be a sequence of discrete functions such that*

$$\lim_{h \to 0} \|f - f_h\|_{H^{-1}} = 0.$$

Then

$$\lim_{h \to 0} \left\|(-\partial_{xx})^{-1} f - (-\Delta_h)^{-1} f_h\right\|_{H_0^1} = 0. \tag{4.18}$$

Besides, if $f \in L^2(0,1)$ and f_h satisfies, for some $\theta > 0$,

$$\|f - f_h\|_{H^{-1}} \leq C_0 h^\theta,$$

then

$$\left\|(-\partial_{xx})^{-1} f - (-\Delta_h)^{-1} f_h\right\|_{H_0^1} \leq C \left(h \|f\|_{L^2} + C_0 h^\theta\right). \tag{4.19}$$

Proof. The first part of Theorem 4.2 easily follows by the density of $L^2(0,1)$ functions in $H^{-1}(0,1)$, the uniform stability result of Proposition 4.1 and the convergence result of Theorem 4.1, similarly as in the proof of Proposition 3.5. The details are left to the reader.

The second part of Theorem 4.2 consists of taking $\tilde{f}_h$ as in Eq. (4.11), for which we have

$$\left\|f - \tilde{f}_h\right\|_{H^{-1}} \le Ch\|f\|_{L^2} \quad \text{and} \quad \left\|(-\Delta_h)^{-1}\tilde{f}_h - (-\partial_{xx})^{-1}f\right\|_{H_0^1} \le Ch\|f\|_{L^2}.$$

Then Proposition 4.1 implies that

$$\left\|(-\Delta_h)^{-1}f_h - (-\Delta_h)^{-1}\tilde{f}_h\right\|_{H_0^1} \le C\left\|f_h - \tilde{f}_h\right\|_{H^{-1}}.$$

Of course, these three last estimates imply Eq. (4.19).   □

Finally, we mention this last result:

Theorem 4.3. *Let $f \in L^2(0,1)$ and $z = (-\partial_{xx})^{-1}f$. Then there exists C such that*

$$|\partial_x z(1)|^2 \le C\|f\|_{L^2}\|f\|_{H^{-1}}. \tag{4.20}$$

Similarly, there exists $C > 0$ such that for all $h \in (0,1)$, if f_h is a discrete function and $z_h = (-\Delta_h)^{-1}f_h$, we have

$$\left|\frac{z_{N,h}}{h}\right|^2 \le C\|f_h\|_{L^2}\|f_h\|_{H^{-1}}. \tag{4.21}$$

Besides, taking f_h as in Eq. (4.11), we have

$$\left|\partial_x z(1) + \frac{z_{N,h}}{h}\right| \le C\sqrt{h}\|f\|_{L^2}. \tag{4.22}$$

Proof. We prove this result using the multiplier technique. Since $-\partial_{xx}z = f$, multiplying the equation by $x\partial_x z$, easy integrations by parts show

$$|\partial_x z(1)|^2 = -2\int_0^1 fx\partial_x z + \int_0^1 |\partial_x z|^2.$$

Of course, this implies Eq. (4.20) from the fact that $\|z\|_{H_0^1} = \|f\|_{H^{-1}}$.

In order to prove estimate (4.21), we develop a similar multiplier argument. Namely, we multiply the equation

$$-(\Delta_h z_h)_j = f_{j,h}, \quad j \in \{1, \dots, N\},$$

by $j(z_{j+1,h} - z_{j-1,h})$. We thus obtain

$$\left|\frac{z_{N,h}}{h}\right|^2 = -2h\sum_{j=1}^N jh\left(\frac{z_{j+1,h} - z_{j-1,h}}{h}\right)f_{j,h} + h\sum_{j=0}^N \left(\frac{z_{j+1,h} - z_{j,h}}{h}\right)^2.$$

Hence

$$\left|\frac{z_{N,h}}{h}\right|^2 \leq C\|f_h\|_{L^2}\|z_h\|_{H_0^1} + C\|z_h\|_{H_0^1}^2 \leq C\|f_h\|_{L^2}\|f_h\|_{H^{-1}} + C\|f_h\|_{H^{-1}}^2,$$

which yields estimate (4.21).

We now aim at proving Eq. (4.22). First remark that z_h also solves

$$-\partial_{xx}z_h = \tilde{f}_h, \text{ on } (0,1), \quad z_h(0) = z_h(1) = 0,$$

where

$$\tilde{f}_h = \sum_{j=1}^{N} f_k \left(\frac{\mu_k}{\mu_k(h)}\right)^2 w^k. \tag{4.23}$$

But one easily has

$$\|\tilde{f}_h\|_{L^2} \leq C\|f\|_{L^2}, \quad \|\tilde{f}_h - f\|_{H^{-1}} \leq Ch\|f\|_{L^2}. \tag{4.24}$$

Indeed, from Eq. (4.17),

$$\|\tilde{f}_h - f_h\|_{H^{-1}}^2 = \sum_{k=1}^{N} \frac{|f_k|^2}{\mu_k^2}\left(1 - \left(\frac{\mu_k}{\mu_k(h)}\right)^2\right)^2 \leq Ch^2\|f\|_{L^2}^2,$$

and thus Eq. (4.14) yields Eq. (4.24).

Therefore, using Eq. (4.21),

$$|\partial_x z(1) - \partial_x z_h(1)| \leq C\left(\|f - \tilde{f}_h\|_{L^2}\|f - \tilde{f}_h\|_{H^{-1}}\right)^{1/2} \leq C\sqrt{h}\|f\|_{L^2}. \tag{4.25}$$

Besides,

$$\partial_x z_h(1) + \frac{z_{N,h}}{h} = \sum_{k=1}^{N} \frac{f_k}{\mu_k(h)^2}(-1)^k\left(1 - \frac{\sin(k\pi h)}{k\pi h}\right)k\pi.$$

Note that this last expression coincides with the continuous normal derivative $\partial_x \tilde{z}(1)$ of the solution $\tilde{z}$ of the continuous problem

$$\begin{cases} -\partial_{xx}\tilde{z} = \tilde{g}_h, \text{ on } (0,1), \text{ where } \tilde{g}_h = \sum_{k=1}^{N} f_k \frac{\mu_k^2}{\mu_k(h)^2}\left(1 - \frac{\sin(k\pi h)}{k\pi h}\right)w^k, \\ \tilde{z}(0) = \tilde{z}(1) = 0. \end{cases} \tag{4.26}$$

Using that for some constant C independent of h and $k \in \{1, \ldots, N\}$,

$$\left|\frac{\mu_k^2}{\mu_k(h)^2}\right| \leq C, \quad \left|1 - \frac{\sin(k\pi h)}{k\pi h}\right| \leq Ck^2 h^2,$$

we easily compute

$$\|\tilde{g}_h\|_{L^2} \leq C \|f\|_{L^2}, \qquad \|\tilde{g}_h\|_{H^{-1}} \leq Ch \|f\|_{L^2}. \tag{4.27}$$

Hence, from Eq. (4.20),

$$\left| \partial_x z_h(1) + \frac{z_{N,h}}{h} \right| = |\partial_x \tilde{z}(1)| \leq C\sqrt{h} \|f\|_{L^2}.$$

Together with Eq. (4.25), this concludes the proof of Theorem 4.3.   □

4.2.2 Stronger Norms

Recalling the definition of the functional spaces $H_{(0)}^\ell(0,1)$ in Eq. (3.34), we prove the counterparts of the above theorem within these spaces.

First, Proposition 4.1 can be modified into:

Proposition 4.2. *Let $\ell \in \mathbb{R}$. If f_h is a discrete function, then there exists a constant $C = C(\ell)$ independent of $h \in (0,1)$ such that*

$$\frac{1}{C} \|f_h\|_{H_{(0)}^\ell} \leq \left\| (-\Delta_h)^{-1} f_h \right\|_{H_{(0)}^{\ell-2}} \leq C \|f_h\|_{H_{(0)}^\ell}. \tag{4.28}$$

The proof of Proposition 4.2 follows line to line the one of Proposition 4.1 and is left to the reader.

The convergence results of Theorem 4.1 can be extended as follows:

Theorem 4.4. *Let $\ell \in \mathbb{R}$ and $f \in H_{(0)}^\ell(0,1)$ and $z = (-\partial_{xx})^{-1} f$ be the corresponding solution of the Laplace equation (4.12). With the notations of Theorem 4.1, setting f_h as in Eq. (4.11) and $z_h = (-\Delta_h)^{-1} f_h$, we have*

$$\|f - f_h\|_{H_{(0)}^{\ell-1}} + \|z - z_h\|_{H_{(0)}^{\ell+1}} \leq Ch \|f\|_{H_{(0)}^\ell}, \tag{4.29}$$

$$\|z - z_h\|_{H_{(0)}^\ell} \leq Ch^2 \|f\|_{H_{(0)}^\ell}. \tag{4.30}$$

Here again, the proof of Theorem 4.4 is very similar to the one of Theorem 4.1 and is left to the reader.

We now focus on the convergence of the normal derivatives:

Theorem 4.5. *Let $\ell \geq 0$ and $f \in H_{(0)}^\ell(0,1)$ and $z = (-\partial_{xx})^{-1} f$ be the corresponding solution of the Laplace equation (4.12). With the notations of Theorem 4.1, setting f_h as in Eq. (4.11) and $z_h = (-\Delta_h)^{-1} f_h$, we have*

$$\left| \partial_x z(1) + \frac{z_{N,h}}{h} \right| \leq Ch^{\min\{\ell+1/2, \ell/2+1, 2\}} \|f\|_{H_{(0)}^\ell}. \tag{4.31}$$

Proof. The proof of Eq. (4.31) follows the one of Eq. (4.22), except for the estimates (4.24) on $\tilde{f}_h$ in Eqs. (4.23) and (4.27) on $\tilde{g}_h$ defined in Eq. (4.26).

Using that for all $h > 0$ and $k \in \{1, \ldots, N\}$,

$$\left(1 - \left(\frac{\mu_k}{\mu_k(h)}\right)^2\right)^2 \leq Ck^4h^4,$$

we easily derive that

$$\|f - \tilde{f}_h\|_{L^2}^2 \leq C\left(\frac{1}{N^{2\ell}} + Ch^4 \max\{1, N^{4-2\ell}\}\right) \|f\|_{H_{(0)}^\ell}^2.$$

In particular, if $\ell \in (0, 2]$, $\|f - \tilde{f}_h\|_{L^2} \leq Ch^\ell \|f\|_{H_{(0)}^\ell}$ and if $\ell \geq 2$, $\|f - \tilde{f}_h\|_{L^2} \leq Ch^2 \|f\|_{H_{(0)}^\ell}$, thus yielding

$$\|f - \tilde{f}_h\|_{L^2} \leq Ch^{\min\{\ell, 2\}} \|f\|_{H_{(0)}^\ell}.$$

Similarly,

$$\|f - \tilde{f}_h\|_{H^{-1}} \leq Ch^{\min\{\ell+1, 2\}} \|f\|_{H_{(0)}^\ell}.$$

We thus obtain, instead of Eq. (4.25),

$$|\partial_x z(1) - \partial_x z_h(1)| \leq Ch^{\min\{\ell+1/2, \ell/2+1, 2\}} \|f\|_{H_{(0)}^\ell}.$$

Estimates on $\partial_x z_h(1) + z_{N,h}/h$ can be deduced similarly from estimates on $\tilde{g}_h$ (defined in Eq. (4.26)) and are left to the reader. □

Remark 4.2. Very likely, estimate (4.31) can be improved for $\ell > -1/2$ into

$$\left|\partial_x z(1) + \frac{z_{N,h}}{h}\right| \leq Ch^{\min\{\ell+1/2, 2\}} \|f\|_{H_{(0)}^\ell}. \tag{4.32}$$

For instance, using that, if $f = \sum_k f_k w^k$, the solution z of Eq. (4.12) can be expanded as $z = \sum_k f_k/\mu_k^2 w^k$ and we get

$$\partial_x z(1) = \sum_k f_k \frac{\partial_x w^k(1)}{\mu_k^2},$$

provided the sum converges. Since for all $k \in \mathbb{N}$,

$$\left|\frac{\partial_x w^k(1)}{\mu_k^2}\right| \leq \frac{C}{\mu_k},$$

by Cauchy–Schwarz, for any $\ell_0 > -1/2$, we obtain

$$|\partial_x z(1)| \le C_{\ell_0} \|f\|_{H^{\ell_0}_{(0)}}$$

instead of Eq. (4.20).

Of course, we can get similar estimates for the discrete solutions $z_h = (-\Delta_h)^{-1} f_h$ and obtain, for all $\ell_{(0)} > -1/2$, a constant C_{ℓ_0} independent of $h > 0$ such that for all discrete function f_h and $z_h = (-\Delta_h)^{-1} f_h$,

$$\left| \frac{z_{N,h}}{h} \right| \le C_{\ell_0} \|f_h\|_{H^{\ell_0}_{(0)}} .$$

instead of Eq. (4.21).

Using these two estimates instead of Eqs. (4.20) and (4.21) and following the proof of Theorem 4.5, we can obtain the following result: for all $\ell > -1/2$ and $\varepsilon > 0$, there exists a constant $C_{\ell,\varepsilon} = C(\ell, \varepsilon)$ such that $f \in H^\ell_{(0)}$,

$$\left| \partial_x z(1) + \frac{z_{N,h}}{h} \right| \le C_{\ell,\varepsilon} h^{\min\{\ell+1/2-\varepsilon, 2\}} \|f\|_{H^\ell_{(0)}} . \tag{4.33}$$

This last estimate is better than Eq. (4.31) when $\ell \in (-1/2, 0)$ and when $\ell \in (1, 2)$.

4.2.3 Numerical Results

This section aims at giving numerical simulations and evidences of the convergence results Eq. (4.31) for the normal derivatives of solutions of the discrete Laplace equation. We do not present a systematic study of the convergence of the solution in $L^2(0,1)$ nor in $H^1_0(0,1)$ since these results are classical and can be found in many textbooks of numerical analysis; see, e.g., [4, 46].

In order to do that, we choose continuous functions f and z solving Eq. (4.12).

For $N \in \mathbb{N}$, we then discretize the source term f into f_h simply by taking $f_h(j) = f(jh)$ for $j \in \{1, \dots, N\}$ and compute z_h the solution of $-\Delta_h z_h = f_h$ with $z_{0,h} = z_{N+1,h} = 0$. We then compute $\partial_x z(1) + z_{N,h}/h$.

Our first test function is

$$f(x) = -\sin(2\pi x) + 3\sin(\pi x), \text{ for } z(x) = \frac{\sin(2\pi x)}{4\pi^2} - \frac{3\sin(\pi x)}{\pi}. \tag{4.34}$$

The plot of $\left| \partial_x z(1) + z_{N,h}/h \right|$ versus N is represented in logarithmic scales in Fig. 4.1, left. Here, we have chosen $N \in [100, 300]$. The slope of the linear regression is -1.99 and completely corresponds to the result of Theorem 4.5.

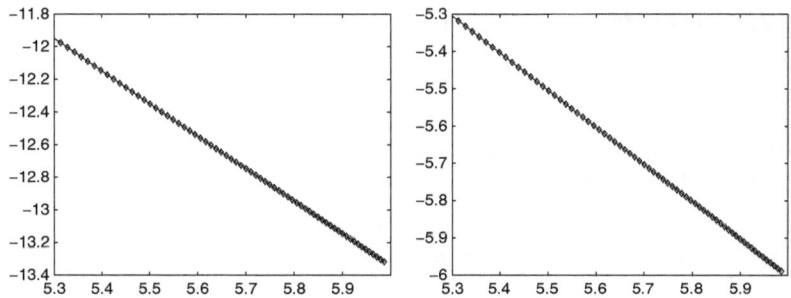

Fig. 4.1 Plot of $|\partial_x z(1) + z_{N,h}/h|$ versus N in logarithmic scales. *Left*, for f as in Eq. (4.34), the slope is -1.99. *Right*, for f as in Eq. (4.35), the slope is -1.00.

We then test

$$f(x) = \frac{1}{(x+1)^3}, \text{ corresponding to } z(x) = -\frac{1}{2(x+1)} + \frac{1}{2} - \frac{x}{4}. \qquad (4.35)$$

Numerical simulations are represented in Fig. 4.1, right.

This function f is smooth, but it does not satisfy $f(0) = f(1) = 0$. Thus it is only in $\cap_{\varepsilon > 0} H_{(0)}^{1/2-\varepsilon}(0,1)$ and the slope predicted by Theorem 4.5 is -1^- and completely agrees with the slope observed in Fig. 4.1 right.

These two examples indicate that the rates of convergence of the normal derivatives obtained in Theorem 4.5 are accurate.

4.3 Uniform Bounds on y_h

The goal of this section is to obtain uniform bounds on y_h in the natural space for the wave equation with nonhomogeneous Dirichlet control, that is $C([0,T];L^2(0,1)) \cap C^1([0,T];H^{-1}(0,1))$:

Theorem 4.6. *There exists a constant C independent of $h > 0$ such that any solution y_h of Eq. (4.7) with initial data (y_h^0, y_h^1) and source term $v_h \in L^2(0,T)$ satisfies*

$$\sup_{t \in [0,T]} \|(y_h(t), \partial_t y_h(t))\|_{L^2(0,1) \times H^{-1}(0,1)}$$

$$\leq C \left(\|(y_h^0, y_h^1)\|_{L^2(0,1) \times H^{-1}(0,1)} + \|v_h\|_{L^2(0,T)} \right). \qquad (4.36)$$

The proof of Theorem 4.6 is done in two steps: one focusing on the estimate on y_h and the other one on $\partial_t y_h$, respectively, corresponding to Propositions 4.3 and 4.4.

As we will see, each one of these propositions is based on a suitable duality argument for solutions of the adjoint system.

4.3.1 Estimates in $C([0,T]; L^2(0,1))$

We have the following:

Proposition 4.3. *There exists a constant C independent of $h > 0$ such that any solution y_h of Eq. (4.7) satisfies*

$$\|y_h\|_{L^\infty(0,T;L^2(0,1))} \leq C \left(\|y_h^0\|_{L^2(0,1)} + \|y_h^1\|_{H^{-1}(0,1)} + \|v_h\|_{L^2(0,T)} \right). \qquad (4.37)$$

We postpone the proof to the end of the section. As in the continuous case, Proposition 4.3 will be a consequence of a suitable duality argument.

Namely, let $f_h \in L^1(0,T;L^2(0,1))$ and define ϕ_h as being the solution of

$$\begin{cases} \partial_{tt}\phi_{j,h} - \dfrac{1}{h^2}\left[\phi_{j+1,h} + \phi_{j-1,h} - 2\phi_{j,h}\right] = f_{j,h}, \\ \qquad\qquad\qquad\qquad (t,j) \in (0,T) \times \{1,\dots,N\}, \\ \phi_{0,h}(t) = \phi_{N+1,h}(t) = 0, \qquad t \in (0,T), \\ \phi_{j,h}(T) = 0, \partial_t\phi_{j,h}(T) = 0, \qquad j = 1,\dots,N. \end{cases} \qquad (4.38)$$

Then, multiplying Eq. (4.7) by ϕ_h solution of Eq. (4.38), we obtain

$$0 = h\sum_{j=1}^N \int_0^T \partial_{tt}y_{j,h}\phi_{j,h}\,dt - h\sum_{j=1}^N \int_0^T \frac{1}{h^2}[y_{j+1,h}+y_{j-1,h}-2y_{j,h}]\phi_{j,h}\,dt$$

$$= h\sum_{j=1}^N \int_0^T y_{j,h}\partial_{tt}\phi_{j,h}\,dt - h\sum_{j=1}^N \int_0^T \frac{1}{h^2}y_{j,h}[\phi_{j+1,h}+\phi_{j-1,h}-2\phi_{j,h}]\,dt$$

$$+ h\sum_{j=1}^N (\partial_t y_{j,h}\phi_{j,h} - y_{j,h}\partial_t\phi_{j,h})\Big|_0^T - \int_0^T y_{N+1,h}\frac{\phi_{N,h}}{h}\,dt$$

$$= h\sum_{j=1}^N \int_0^T y_{j,h}f_{j,h}\,dt + h\sum_{j=1}^N (y_{j,h}^0\partial_t\phi_{j,h}(0) - y_{j,h}^1\phi_{j,h}(0)) \qquad (4.39)$$

$$- \int_0^T v_h(t)\frac{\phi_{N,h}(t)}{h}\,dt.$$

Note that identity (4.39) is a discrete counterpart of the continuous identity (4.4). Remark that this can be used as a definition of solutions of Eq. (4.7) by transposition, even if in that case, solutions of Eq. (4.7) obviously exist due to the finite dimensional nature of system (4.7).

Formulation (4.39) will be used to derive estimates on solutions y_h by duality.

But we shall first prove the following lemma:

Lemma 4.1. *For ϕ_h solution of Eq. (4.38), there exists a constant C independent of $h > 0$ such that*

$$\|\phi_h\|_{L^\infty(0,T;H_0^1(0,1))} + \|\partial_t\phi_h\|_{L^\infty(0,T;L^2(0,1))} \leq C\|f_h\|_{L^1(0,T;L^2(0,1))} \qquad (4.40)$$

and

$$\left\|\frac{\phi_{N,h}}{h}\right\|_{L^2(0,T)} \leq C\|f_h\|_{L^1(0,T;L^2(0,1))}. \tag{4.41}$$

Proof. The first inequality (4.40) is an energy estimate, whereas Eq. (4.41) is a hidden regularity property.

Multiplying Eq. (4.38) by $\partial_t \phi_{j,h}$ and summing over j, we obtain

$$h\sum_{j=1}^{N} \partial_{tt}\phi_{j,h}\partial_t\phi_{j,h} - h\sum_{j=1}^{N}\frac{1}{h^2}\left[\phi_{j+1,h}+\phi_{j-1,h}-2\phi_{j,h}\right]\partial_t\phi_{j,h}$$

$$= h\sum_{j=1}^{N} f_{j,h}\partial_t\phi_{j,h}. \tag{4.42}$$

The left-hand side of Eq. (4.42) is the derivative of the energy

$$\frac{\mathrm{d}}{\mathrm{d}t}\left(\frac{h}{2}\sum_{j=1}^{N}|\partial_t\phi_{j,h}|^2 + \frac{h}{2}\sum_{j=1}^{N}\left(\frac{\phi_{j+1,h}-\phi_{j,h}}{h}\right)^2\right) = \frac{1}{2}\frac{\mathrm{d}E_h[\phi_h]}{\mathrm{d}t},$$

whereas the right-hand side satisfies

$$\left|h\sum_{j=1}^{N} f_{j,h}\partial_t\phi_{j,h}\right| \leq \left(h\sum_{j=1}^{N}|f_{j,h}|^2\right)^{1/2}\left(h\sum_{j=1}^{N}|\partial_t\phi_{j,h}|^2\right)^{1/2}$$

$$\leq \left(h\sum_{j=1}^{N}|f_{j,h}|^2\right)^{1/2}\sqrt{E_h[\phi_h](t)}.$$

Equation (4.42) then implies

$$\left|\frac{\mathrm{d}\sqrt{E_h}}{\mathrm{d}t}(t)\right| \leq \left(h\sum_{j=1}^{N}|f_{j,h}(t)|^2\right)^{1/2}. \tag{4.43}$$

Integrating in time, we obtain that for all $t \in [0,T]$,

$$\sqrt{E_h(t)} \leq \int_0^T \left(h\sum_{j=1}^{N}|f_{j,h}(t)|^2\right)^{1/2}\mathrm{d}t.$$

Finally, recalling the properties of the Fourier extension operator in Sect. 3.2, we obtain Eq. (4.40).

Estimate (4.41) can be deduced from the multiplier approach developed in the proof of Theorem 2.2 by multiplying Eq. (4.38) by $j(\phi_{j+1,h} - \phi_{j-1,h})$:

$$h \sum_{j=1}^{N} \int_0^T f_{j,h} \, jh \left(\frac{\phi_{j+1,h} - \phi_{j-1,h}}{h} \right) dt$$

$$= h \sum_{j=1}^{N} \int_0^T \partial_{tt} \phi_{j,h} \, jh \left(\frac{\phi_{j+1,h} - \phi_{j-1,h}}{h} \right) dt$$

$$- h \sum_{j=1}^{N} \int_0^T \left[\frac{\phi_{j+1,h} + \phi_{j-1,h} - 2\phi_{j,h}}{h^2} \right] jh \left(\frac{\phi_{j+1,h} - \phi_{j-1,h}}{h} \right) dt. \quad (4.44)$$

The right-hand side of Eq. (4.44) has already been dealt with in the proof of Theorem 2.2 and yields

$$h \sum_{j=1}^{N} \int_0^T \partial_{tt} \phi_{j,h} \, jh \left(\frac{\phi_{j+1,h} - \phi_{j-1,h}}{h} \right) dt$$

$$- h \sum_{j=1}^{N} \int_0^T \left[\frac{\phi_{j+1,h} + \phi_{j-1,h} - 2\phi_{j,h}}{h^2} \right] jh \left(\frac{\phi_{j+1,h} - \phi_{j-1,h}}{h} \right)$$

$$= \int_0^T \left| \frac{\phi_{N,h}(t)}{h} \right|^2 dt + \frac{h^3}{2} \sum_{j=0}^{N} \int_0^T \left| \frac{\partial_t \phi_{j+1,h} - \partial_t \phi_{j,h}}{h} \right|^2 dt$$

$$- \int_0^T E_h(t) \, dt - X_h(t) \Big|_0^T,$$

where, similarly as in Eq. (2.14), $X_h(t)$ is given by

$$X_h(t) = 2h \sum_{j=1}^{N} jh \left(\frac{\phi_{j+1,h} - \phi_{j-1,h}}{2h} \right) \partial_t \phi_{j,h}.$$

From the conditions $\phi_h(T) = \partial_t \phi_h(T) = 0$ in Eq. (4.38), $X_h(T) = 0$. Besides, as in Eq. (2.15), one has $|X_h(0)| \leq E_h(0)$.

On the other hand,

$$\left| h \sum_{j=1}^{N} \int_0^T f_{j,h} \, jh \left(\frac{\phi_{j+1,h} - \phi_{j-1,h}}{h} \right) dt \right|$$

$$\leq \int_0^T \left(h \sum_{j=1}^{N} |f_{j,h}|^2 \right)^{1/2} \sqrt{E_h(t)} \, dt$$

$$\leq \sup_{t \in [0,T]} \left\{ \sqrt{E_h(t)} \right\} \int_0^T \left(h \sum_{j=1}^{N} |f_{j,h}|^2 \right)^{1/2} dt.$$

Therefore, from Eq. (4.40), there exists a constant independent of h such that

$$\int_0^T \left| \frac{\phi_{N,h}(t)}{h} \right|^2 dt + \frac{h^3}{2} \sum_{j=0}^N \int_0^T \left| \frac{\partial_t \phi_{j+1,h} - \partial_t \phi_{j,h}}{h} \right|^2 dt$$

$$\leq C \left(\int_0^T \left(h \sum_{j=1}^N |f_{j,h}|^2 \right)^{1/2} dt \right)^2,$$

which implies Eq. (4.41). □

Proof (Proposition 4.3). Lemma 4.1 and identity (4.39) allow us to deduce bounds on y_h. Indeed,

$$\|y_h\|_{L^\infty(0,T;L^2(0,1))} = \sup_{\substack{f \in L^1(0,T;L^2(0,1)) \\ \|f\|_{L^1((0,T);L^2(0,1))}}} \int_0^1 y_h(x) f(x) \, dx.$$

But there y_h is the Fourier extension $\mathbb{F}_h(y_h)$ (recall Sect. 3.2); hence it involves only Fourier modes smaller than N. We thus only have to consider the projection of f onto the first N Fourier modes. But this exactly corresponds to discrete functions f_h. Therefore,

$$\|y_h\|_{L^\infty(0,T;L^2(0,1))} = \sup_{\substack{f_h \in L^1(0,T;L^2(0,1)) \\ \|f_h\|_{L^1((0,T);L^2(0,1))} \leq 1}} \left\{ h \sum_{j=1}^N \int_0^T y_{j,h} f_{j,h} \, dt \right\}.$$

But, introducing ϕ_h, the solution of Eq. (4.38) with source term f_h, using Lemma 4.1, we obtain:

$$h \sum_{j=1}^N \int_0^T y_{j,h} f_{j,h} \, dt = -h \sum_{j=1}^N (y_{j,h}^0 \partial_t \phi_{j,h}(0) - y_{j,h}^1 \phi_{j,h}(0)) + \int_0^T v_h(t) \frac{\phi_{N,h}(t)}{h} \, dt$$

$$\leq C \|y_h^0\|_{L^2(0,1)} \|\partial_t \phi_h(0)\|_{L^2(0,1)} + C \|y_h^1\|_{H^{-1}(0,1)} \|\phi_h(0)\|_{H_0^1(0,1)}$$

$$+ \|v_h\|_{L^2(0,T)} \left\| \frac{\phi_{N,h}}{h} \right\|_{L^2(0,T)}$$

$$\leq C \left(\|y_h^0\|_{L^2(0,1)} + \|y_h^1\|_{H^{-1}(0,1)} + \|v_h\|_{L^2(0,T)} \right) \|f_h\|_{L^1(0,T;L^2(0,1))}.$$

This yields in particular Eq. (4.37). □

4.3.2 Estimates on $\partial_t y_h$

We now focus on getting estimates on $\partial_t y_h$.

Proposition 4.4. *There exists a constant C independent of $h > 0$ such that any solution y_h of Eq. (4.7) satisfies*

$$\|\partial_t y_h\|_{L^\infty(0,T;H^{-1}(0,1))} \le C\left(\|y_h^0\|_{L^2(0,1)} + \|y_h^1\|_{H^{-1}(0,1)} + \|v_h\|_{L^2(0,T)}\right). \quad (4.45)$$

Similarly as for Proposition 4.3, this result is obtained by duality, based on the following identity: if ϕ_h solves the adjoint wave equation (4.38) with source term $f_h = \partial_t g_h$ with $g_h \in L^1(0,T;H_0^1(0,1))$, we have:

$$h\sum_{j=1}^N \int_0^T y_{j,h}\partial_t g_{j,h}\,dt = -h\sum_{j=1}^N (y_{j,h}^0 \partial_t \phi_{j,h}(0) - y_{j,h}^1 \phi_{j,h}(0))$$

$$+ \int_0^T v_h(t)\frac{\phi_{N,h}(t)}{h}\,dt. \quad (4.46)$$

The proof of Proposition 4.4 is sketched at the end of the section, since it is very similar to the one of Proposition 4.3.

Hence, we focus on the following adjoint problem:

$$\begin{cases} \partial_{tt}\phi_{j,h} - \dfrac{1}{h^2}\left[\phi_{j+1,h} + \phi_{j-1,h} - 2\phi_{j,h}\right] = \partial_t g_{j,h}, \\ \hspace{4cm} (t,j) \in (0,T) \times \{1,\ldots,N\}, \\ \phi_{0,h}(t) = \phi_{N+1,h}(t) = 0, \hspace{1.3cm} t \in (0,T), \\ \phi_{j,h}(T) = 0, \partial_t \phi_{j,0}(T) = 0, \hspace{0.9cm} j = 1,\ldots,N. \end{cases} \quad (4.47)$$

We shall thus prove the following:

Lemma 4.2. *For ϕ_h solution of Eq. (4.47), there exists a constant C independent of $h > 0$ such that*

$$\|\phi_h\|_{L^\infty(0,T;H_0^1(0,1))} + \|\partial_t \phi_h(0)\|_{L^2(0,1)} \le C\|g_h\|_{L^1(0,T;H_0^1(0,1))} \quad (4.48)$$

and

$$\left\|\frac{\phi_{N,h}}{h}\right\|_{L^2(0,T)} \le C\|g_h\|_{L^1(0,T;H_0^1(0,1))}. \quad (4.49)$$

Proof. To study solutions ϕ_h of Eq. (4.47), it is convenient to first assume that g_h is compactly supported in time in $(0,T)$ and use the density of compactly supported functions in time in $L^1(0,T;H_0^1(0,1))$.

Let us introduce ψ_h satisfying $\partial_t \psi_h = \phi_h$, which satisfies

$$\begin{cases} \partial_{tt}\psi_{j,h} - \dfrac{1}{h^2}\left[\psi_{j+1,h} + \psi_{j-1,h} - 2\psi_{j,h}\right] = g_{j,h}, \\ \hspace{4cm} (t,j) \in (0,T) \times \{1,\ldots,N\}, \\ \psi_{0,h}(t) = \psi_{N+1,h}(t) = 0, \hspace{1.3cm} t \in (0,T), \\ \psi_{j,h}(T) = 0, \partial_t \psi_{j,h}(T) = 0, \hspace{0.9cm} j = 1,\ldots,N. \end{cases} \quad (4.50)$$

Obviously, using Lemma 4.1, we immediately obtain

$$\|\psi_h\|_{L^\infty(0,T;H^1_0(0,1))} + \|\partial_t\psi_h\|_{L^\infty(0,T;L^2(0,1))} + \left\|\frac{\psi_{N,h}}{h}\right\|_{L^2(0,T)} \le C\|g_h\|_{L^1(0,T;L^2(0,1))}$$
$$\le C\|g_h\|_{L^1(0,T;H^1_0(0,1))}.$$

To derive more precise estimates on ϕ_h, we multiply Eq. (4.50) by $-(\partial_t\psi_{j+1,h} + \partial_t\psi_{j-1,h} - 2\partial_t\psi_{j,h})/h^2$:

$$\frac{d}{dt}\left(\frac{h}{2}\sum_{j=0}^{N}\left(\frac{\partial_t\psi_{j+1,h}-\partial_t\psi_{j,h}}{h}\right)^2 + \frac{h}{2}\sum_{j=1}^{N}\left(\frac{\psi_{j+1,h}+\psi_{j-1,h}-2\psi_{j,h}}{h^2}\right)^2\right)$$
$$= h\sum_{j=1}^{N}\left(\frac{g_{j+1,h}-g_{j,h}}{h}\right)\left(\frac{\partial_t\psi_{j+1,h}-\partial_t\psi_{j,h}}{h}\right).$$

Arguing as in Eq. (4.43), this allows to conclude that

$$\sup_{t\in[0,T]}\left\{\frac{h}{2}\sum_{j=0}^{N}\left(\frac{\partial_t\psi_{j+1,h}-\partial_t\psi_{j,h}}{h}\right)^2 + \frac{h}{2}\sum_{j=1}^{N}\left(\frac{\psi_{j+1,h}+\psi_{j-1,h}-2\psi_{j,h}}{h^2}\right)^2\right\}$$
$$\le C\left(\int_0^T\left(h\sum_{j=0}^{N}\left(\frac{g_{j+1,h}-g_{j,h}}{h}\right)^2\right)^{1/2}dt\right)^2. \tag{4.51}$$

Using Eq. (4.38) and $\partial_t\psi_h = \phi_h$ and again the equivalences proven in Sect. 3.2, we deduce

$$\|\phi_h\|_{L^\infty(0,T;\,H^1_0(0,1))} + \|\partial_{tt}\psi_h + g_h\|_{L^\infty((0,T);L^2(0,1))} \le C\|g_h\|_{L^1(0,T;H^1_0(0,1))},$$

where we use the equation of ψ_h. In order to get Eq. (4.48), we only use the fact that $g_h(0) = 0$.

To deduce Eq. (4.49), we need to apply a multiplier technique on the Eq. (4.47) directly.

Multiplying Eq. (4.47) by $j(\phi_{j+1,h} - \phi_{j-1,h})$, we obtain, similarly as in Eq. (2.13),

$$\int_0^T\left|\frac{\phi_{N,h}(t)}{h}\right|^2 dt + \frac{h^3}{2}\sum_{j=0}^{N}\int_0^T\left|\frac{\partial_t\phi_{j+1,h}-\partial_t\phi_{j,h}}{h}\right|^2 dt$$
$$= \int_0^T E_h(t)\,dt - X_h(0) - h\int_0^T\sum_{j=1}^{N}jh\left(\frac{\phi_{j+1,h}-\phi_{j-1,h}}{h}\right)\partial_t g_{j,h}\,dt, \tag{4.52}$$

where X_h is as in Eq. (2.14). To derive Eq. (4.49), it is then sufficient to bound each term in the right-hand side of this identity.

First remark that

$$\int_0^T E_h(t)\,dt = h\int_0^T \sum_{j=0}^N \left(\frac{\phi_{j+1,h}-\phi_{j,h}}{h}\right)^2 dt + h\int_0^T \sum_{j=0}^N |\partial_t \phi_{j,h}|^2\,dt$$

$$= h\int_0^T \sum_{j=0}^N \left(\frac{\partial_t \psi_{j+1,h}-\partial_t \psi_{j,h}}{h}\right)^2 dt + h\int_0^T \sum_{j=0}^N |\partial_{tt}\psi_{j,h}|^2\,dt$$

$$= h\int_0^T \sum_{j=0}^N \left(\frac{\partial_t \psi_{j+1,h}-\partial_t \psi_{j,h}}{h}\right)^2 dt + h\int_0^T \sum_{j=1}^N \left(\frac{\psi_{j+1,h}+\psi_{j-1,h}-2\psi_{j,h}}{h^2}\right)^2 dt$$

$$+ h\int_0^T \sum_{j=0}^N g_{j,h}^2\,dt + 2h\int_0^T \sum_{j=1}^N \left(\frac{\psi_{j+1,h}+\psi_{j-1,h}-2\psi_{j,h}}{h^2}\right)g_{j,h}\,dt.$$

In particular, from Eq. (4.51), we obtain

$$\left|\int_0^T E_h(t)\,dt - h\int_0^T \sum_{j=0}^N g_{j,h}^2\,dt\right| \le C\|g\|_{L^1(0,T;H_0^1(0,1))}^2.$$

Let us then bound $X_h(0)$. Since $g_h(0)=0$,

$$X_h(0) = 2h\sum_{j=1}^N jh\left(\frac{\phi_{j+1,h}(0)-\phi_{j-1,h}(0)}{2h}\right)\partial_t \phi_j(0)$$

$$= 2h\sum_{j=1}^N jh\left(\frac{\phi_{j+1,h}(0)-\phi_{j-1,h}(0)}{2h}\right)\partial_{tt}\psi_j(0)$$

$$= 2h\sum_{j=1}^N jh\left(\frac{\phi_{j+1,h}(0)-\phi_{j-1,h}(0)}{2h}\right)\left(\frac{\psi_{j+1,h}(0)+\psi_{j-1,h}(0)-2\psi_{j,h}(0)}{h^2}\right).$$

It follows then from Eq. (4.51) that

$$|X_h(0)| \le C\|g_h\|_{L^1(0,T;H_0^1(0,1))}^2.$$

We now deal with the last term in Eq. (4.52):

$$I := 2h\int_0^T \sum_{j=1}^N jh\left(\frac{\phi_{j+1,h}-\phi_{j-1,h}}{2h}\right)\partial_t g_{j,h}\,dt.$$

Integrating by parts we get

$$I = -h\int_0^T \sum_{j=1}^N \phi_{j,h}\left((j+1)\partial_t g_{j+1,h}-(j-1)\partial_t g_{j-1,h}\right)dt$$

$$= -h\int_0^T \sum_{j=1}^N \phi_{j,h}\left((\partial_t g_{j-1,h}+\partial_t g_{j+1,h})+jh\left(\frac{\partial_t g_{j+1,h}-\partial_t g_{j-1,h}}{h}\right)\right)dt.$$

Taking into account that, by assumption, $g_h(0) = g_h(T) = 0$,

$$I = h \int_0^T \sum_{j=1}^N \partial_t \phi_{j,h} \left((g_{j-1,h} + g_{j+1,h}) + jh \left(\frac{g_{j+1,h} - g_{j-1,h}}{h} \right) \right) dt.$$

But $\partial_t \phi_{j,h} = \partial_{tt} \psi_{j,h}$, and then Eq. (4.50) yields:

$$I = h \int_0^T \sum_{j=1}^N g_{j,h} \left((g_{j-1,h} + g_{j+1,h}) + jh \left(\frac{g_{j+1,h} - g_{j-1,h}}{h} \right) \right) dt$$

$$+ h \int_0^T \sum_{j=1}^N \left(\frac{\psi_{j+1,h} + \psi_{j-1,h} - 2\psi_{j,h}}{h^2} \right) (g_{j-1,h} + g_{j+1,h}) \, dt.$$

$$+ h \int_0^T \sum_{j=1}^N \left(\frac{\psi_{j+1,h} + \psi_{j-1,h} - 2\psi_{j,h}}{h^2} \right) jh \left(\frac{g_{j+1,h} - g_{j-1,h}}{h} \right) dt.$$

Since

$$h \int_0^T \sum_{j=1}^N g_{j,h} \left((g_{j-1,h} + g_{j+1,h}) + jh \left(\frac{g_{j+1,h} - g_{j-1,h}}{h} \right) \right) dt$$

$$= h \int_0^T \sum_{j=1}^N g_{j,h} g_{j+1,h} \, dt,$$

due to estimates (4.51), we obtain

$$\left| I - h \int_0^T \sum_{j=1}^N g_{j,h} g_{j+1,h} \, dt \right| \leq C \|g\|_{L^1(0,T;H_0^1(0,1))}^2.$$

These estimates, combined with Eq. (4.52), finally give

$$\left| \int_0^T \left| \frac{\phi_{N,h}(t)}{h} \right|^2 dt + \frac{h^3}{2} \sum_{j=0}^N \int_0^T \left| \frac{\partial_t \phi_{j+1,h} - \partial_t \phi_{j,h}}{h} \right|^2 dt \right.$$

$$\left. - h \int_0^T \sum_{j=1}^N \left(|g_{j,h}|^2 - g_{j,h} g_{j+1,h} \right) dt \right| \leq C \|g\|_{L^1(0,T;H_0^1(0,1))}^2,$$

or, equivalently,

$$\left| \int_0^T \left| \frac{\phi_{N,h}(t)}{h} \right|^2 dt + \frac{h}{2} \sum_{j=0}^N \int_0^T |\partial_t \phi_{j+1,h} - \partial_t \phi_{j,h}|^2 dt \right.$$

$$\left. - \frac{h}{2} \int_0^T \sum_{j=0}^N |g_{j+1,h} - g_{j,h}|^2 dt \right| \leq C \|g\|_{L^1(0,T;H_0^1(0,1))}^2. \qquad (4.53)$$

Remark then that

$$
h\sum_{j=0}^{N}\int_{0}^{T}\left|\partial_{t}\phi_{j+1,h}-\partial_{t}\phi_{j,h}\right|^{2}dt-h\int_{0}^{T}\sum_{j=0}^{N}|g_{j+1,h}-g_{j,h}|^{2}dt
$$

$$
=h\sum_{j=0}^{N}\int_{0}^{T}\left|\partial_{tt}\psi_{j+1,h}-\partial_{tt}\psi_{j,h}\right|^{2}dt-h\int_{0}^{T}\sum_{j=0}^{N}|g_{j+1,h}-g_{j,h}|^{2}dt
$$

$$
=h\sum_{j=0}^{N}\int_{0}^{T}\left(\frac{\psi_{j+2,h}+\psi_{j,h}-2\psi_{j+1,h}}{h^{2}}-\frac{\psi_{j+1,h}+\psi_{j-1,h}-2\psi_{j,h}}{h^{2}}\right)^{2}dt
$$

$$
+2h\sum_{j=0}^{N}\int_{0}^{T}\left(\frac{\psi_{j+2,h}+\psi_{j,h}-2\psi_{j+1,h}}{h^{2}}\right)(g_{j+1,h}-g_{j,h})dt,
$$

$$
-2h\sum_{j=0}^{N}\int_{0}^{T}\left(\frac{\psi_{j+1,h}+\psi_{j-1,h}-2\psi_{j,h}}{h^{2}}\right)(g_{j+1,h}-g_{j,h})dt,
$$

with the notation $\psi_{-1,h}=-\psi_{1,h}$ and $\psi_{N+2,h}=-\psi_{N,h}$.

In view of Eq. (4.51), we have

$$
\left|h\sum_{j=0}^{N}\int_{0}^{T}\left|\partial_{t}\phi_{j+1,h}-\partial_{t}\phi_{j,h}\right|^{2}dt-h\int_{0}^{T}\sum_{j=0}^{N}|g_{j+1,h}-g_{j,h}|^{2}dt\right|
$$

$$
\leq C\|g\|_{L^{1}(0,T;H_{0}^{1}(0,1))}^{2}.
$$

Estimate (4.49) then follows directly from Eq. (4.53). □

Proof (Proposition 4.4). Since y_h is a smooth function of time and space (recall that y_h has been identified with its Fourier extension; see Sect. 3.2),

$$
\|\partial_{t}y_{h}\|_{L^{\infty}((0,T);H^{-1}(0,1))}=\sup_{\substack{g\in L^{1}((0,T);H_{0}^{1}(0,1))\\ \|g\|_{L^{1}((0,T);H_{0}^{1}(0,1))}\leq 1}}\int_{0}^{T}\partial_{t}y_{h}g.
$$

As in the proof of Proposition 4.3, we can take the supremum of the functions $g\in L^{1}(0,T;H_{0}^{1}(0,1))$ that are Fourier extensions of discrete functions. Therefore, using Lemma 4.2 together with the duality identity (4.46), we immediately obtain Proposition 4.4. □

4.4 Convergence Rates for Smooth Data

4.4.1 Main Convergence Result

Our goal is to show the following result:

Theorem 4.7. *Let* $(y^0, y^1) \in H_0^1(0,1) \times L^2(0,1)$ *and* $v \in H^1(0,T)$ *be such that* $v(0) = 0$ *and* y *the corresponding solution of Eq.* (4.1) *with initial data* (y^0, y^1) *and boundary condition* v.

Then there exists a discrete sequence of initial data (y_h^0, y_h^1) *such that the solution* y_h *of Eq.* (4.7) *with initial data* (y_h^0, y_h^1) *and boundary data* v *satisfies the following convergence rates:*

- *Convergence of* y_h: *the following convergence estimates hold:*

$$\sup_{t \in [0,T]} \|y_h(t) - y(t)\|_{L^2} \leq C \left(h^{2/3} \|(y^0, y^1)\|_{H_0^1 \times L^2} + h^{1/2} \|v\|_{H^1} \right). \quad (4.54)$$

If we furthermore assume that $v(T) = 0$,

$$\|y_h(T) - y(T)\|_{L^2} \leq Ch^{2/3} \left(\|(y^0, y^1)\|_{H_0^1 \times L^2} + \|v\|_{H^1} \right). \quad (4.55)$$

- *Convergence of* $\partial_t y_h$: *the following convergence estimates hold:*

$$\sup_{t \in [0,T]} \|\partial_t y_h(t) - \partial_t y(t)\|_{H^{-1}} \leq Ch^{2/3} \left(\|(y^0, y^1)\|_{H_0^1 \times L^2} + \|v\|_{H^1} \right). \quad (4.56)$$

Remark 4.3. The above convergences (4.54) and (4.56) may appear surprising since the rates of convergence of the displacement and of the velocity are not the same except when $v(T) = 0$. We refer to Sect. 4.4.2 for the details of the proof.

More curiously, the rates of convergence for the displacement are not the same depending on the fact that $v(T) = 0$ or not. This definitely is a surprise. In the proof below, we will see that this is due to the rate Eq. (4.22) of convergence of the normal derivative for solutions of the Laplace operator.

The proof is divided in two main steps, namely one focusing on the convergence of y_h towards y and the other one on the convergence of $\partial_t y_h$ to $\partial_t y$, these two estimates being the object of the next sections.

Also, recall that under the assumptions of Theorem 4.7, the solution y of Eq. (4.1) lies in $C([0,T]; H^1(0,1))$, its time derivative $\partial_t y$ belongs to $C([0,T]; L^2(0,1))$ and Δy to $C([0,T]; H^{-1}(0,1))$.

As in the case of homogeneous Dirichlet boundary conditions, we will write down

$$y^0 = \sum_{k=1}^{\infty} \hat{y}_k^0 w^k, \quad y^1 = \sum_{k=1}^{\infty} \hat{y}_k^1 w^k, \quad (4.57)$$

whose $H_0^1(0,1) \times L^2(0,1)$-norm coincides with

$$\|(y^0, y^1)\|_{H_0^1 \times L^2}^2 = \sum_{k=1}^{\infty} k^2 \pi^2 |\hat{y}_k^0|^2 + \sum_{k=1}^{\infty} |\hat{y}_k^1|^2 < \infty.$$

We will then choose the initial data (y_h^0, y_h^1) of the form

$$y_h^0 = \sum_{k=1}^{N} \hat{y}_k^0 w^k, \quad y_h^1 = \sum_{k=1}^{N} \hat{y}_k^1 w^k. \tag{4.58}$$

4.4.2 Convergence of y_h

Proposition 4.5. *Under the assumptions of Theorem 4.7, taking (y_h^0, y_h^1) as in Eq. (4.58), we have the convergences (4.54) and Eq. (4.55).*

Proof. To estimate the convergence of y_h to y at time T, we write

$$\|y_h(T) - y(T)\|_{L^2} = \sup_{\substack{\phi_T \in L^2(0,1) \\ \|\phi_T\|_{L^2(0,1)} \leq 1}} \left\{ \int_0^1 (y_h(T) - y(T)) \phi_T \right\}. \tag{4.59}$$

We thus fix $\phi_T \in L^2(0,1)$ and compute

$$\int_0^1 (y_h(T) - y(T)) \phi_T. \tag{4.60}$$

We expand ϕ_T on its Fourier basis:

$$\phi_T = \sum_{k=1}^{\infty} \hat{\phi}_k w^k, \quad \sum_{k=1}^{\infty} |\hat{\phi}_k|^2 < \infty. \tag{4.61}$$

4.4.2.1 Computation of $\int_0^1 y(T) \phi_T$

Let us now compute $\int_0^1 y(T) \phi_T$. In order to do that, we introduce φ solution of

$$\begin{cases} \partial_{tt} \varphi - \partial_{xx} \varphi = 0, & (t,x) \in (0,T) \times (0,1), \\ \varphi(t,0) = \varphi(t,1) = 0, & t \in (0,T), \\ \varphi(T) = 0, \ \partial_t \varphi(T) = \phi_T. \end{cases} \tag{4.62}$$

Then, multiplying Eq. (4.1) by φ, we easily obtain

$$\int_0^1 y(T) \phi_T = \int_0^T v(t) \partial_x \varphi(t,1) \, dt + \int_0^1 y^0 \partial_t \varphi(0) - \int_0^1 y^1 \varphi(0). \tag{4.63}$$

But $v(t) = \int_0^t \partial_t v(s) \, ds$, thus yielding

$$\int_0^T v(t) \partial_x \varphi(t,1) \, dt = \int_0^T \partial_t v(t) \left(\int_t^T \partial_x \varphi(s,1) \, ds \right) dt.$$

We therefore introduce $\Phi(t) = \int_t^T \varphi(s)\,ds$. One then easily checks that

$$\int_0^1 y(T)\phi_T = \int_0^T \partial_t v(t)\partial_x \Phi(t,1)\,dt - \int_0^1 y^0 \partial_{tt}\Phi(0) + \int_0^1 y^1 \partial_t\Phi(0), \qquad (4.64)$$

where Φ solves

$$\begin{cases} \partial_{tt}\Phi - \partial_{xx}\Phi = -\phi_T, & (t,x) \in (0,T) \times (0,1), \\ \Phi(t,0) = \Phi(t,1) = 0, & t \in (0,T), \\ \Phi(T) = 0,\ \partial_t\Phi(T) = 0. \end{cases} \qquad (4.65)$$

We also introduce z_T the solution of

$$- \partial_{xx} z_T = \phi_T, \quad \text{on } (0,1), \qquad z_T(0) = z_T(1) = 0, \qquad (4.66)$$

so that

$$\Psi = \Phi - z_T \qquad (4.67)$$

satisfies

$$\begin{cases} \partial_{tt}\Psi - \partial_{xx}\Psi = 0, & (t,x) \in (0,T) \times (0,1) \\ \Psi(t,0) = \Psi(t,1) = 0, & t \in (0,T), \\ \Psi(T) = z_T,\ \partial_t\Psi(T) = 0. \end{cases} \qquad (4.68)$$

and

$$\int_0^1 y(T)\phi_T = \int_0^T \partial_t v(t)\partial_x \Psi(t,1)\,dt - \int_0^1 y^0 \partial_{tt}\Psi(0) + \int_0^1 y^1 \partial_t\Psi(0)$$
$$+ \int_0^T \partial_t v(t)\partial_x z_T(1)\,dt,$$

and, using that z_T is independent of time,

$$\int_0^1 y(T)\phi_T = \int_0^T \partial_t v(t)\partial_x \Psi(t,1)\,dt - \int_0^1 y^0 \partial_{tt}\Psi(0) + \int_0^1 y^1 \partial_t\Psi(0)$$
$$+ v(T)\partial_x z_T(1). \qquad (4.69)$$

4.4.2.2 Computation of $\int_0^1 y_h(T)\phi_T$

Expanding $y_h(T)$ in discrete Fourier series, we get

$$\int_0^1 y_h(T)\phi_T = \int_0^1 y_h(T)\phi_{T,h} = h \sum_{j=1}^N y_{j,h}(T)\phi_{j,T,h},$$

where

$$\phi_{j,T,h} = \sum_{k=1}^{N} \hat{\phi}_k w_j^k, \quad j \in \{1,\dots,N\}. \tag{4.70}$$

Then, similarly as in Eq. (4.64), we can prove

$$\int_0^1 y_h(T)\phi_T = -\int_0^T \partial_t v(t) \frac{\Phi_{N,h}}{h} \, dt - h \sum_{j=1}^{N} y_{j,h}^0 \partial_{tt} \Phi_{j,h}(0) + h \sum_{j=1}^{N} y_{j,h}^1 \partial_t \Phi_{j,h}(0), \tag{4.71}$$

where Φ_h is the solution of

$$\begin{cases} \partial_{tt} \Phi_{j,h} - \dfrac{1}{h^2} \left(\Phi_{j+1,h} - 2\Phi_{j,h} + \Phi_{j-1,h} \right) = -\phi_{j,T,h}, \\ \qquad\qquad\qquad\qquad (t,j) \in (0,T) \times \{1,\dots,N\}, \quad (4.72) \\ \Phi_{0,h}(t) = \Phi_{N+1,h}(t) = 0, \qquad t \in (0,T), \\ \Phi_h(T) = 0, \ \partial_t \Phi_h(T) = 0. \end{cases}$$

Note that, due to the orthogonality properties of the Fourier basis, we can write

$$-h \sum_{j=1}^{N} y_{j,h}^0 \partial_{tt} \Phi_{j,h}(0) + h \sum_{j=1}^{N} y_{j,h}^1 \partial_t \Phi_{j,h}(0) = -\int_0^1 y_h^0 \partial_{tt} \Phi_h(0) + \int_0^1 y_h^1 \partial_t \Phi_h(0)$$

$$= -\int_0^1 y^0 \partial_{tt} \Phi_h(0) + \int_0^1 y^1 \partial_t \Phi_h(0),$$

and thus Eq. (4.71) can be rewritten as

$$\int_0^1 y_h(T)\phi_T = -\int_0^T \partial_t v(t) \frac{\Phi_{N,h}}{h} \, dt - \int_0^1 y^0 \partial_{tt} \Phi_h(0) + \int_0^1 y^1 \partial_t \Phi_h(0). \tag{4.73}$$

Then setting

$$z_{T,h} = (-\Delta_h)^{-1} \phi_{T,h}, \tag{4.74}$$

we obtain

$$\int_0^1 y_h(T)\phi_T = -\int_0^T \partial_t v(t) \frac{\Psi_{N,h}}{h} \, dt - \int_0^1 y^0 \partial_{tt} \Psi_h(0) + \int_0^1 y^1 \partial_t \Psi_h(0) \tag{4.75}$$

$$-v(T)\frac{z_{N,T,h}}{h},$$

where Ψ_h is the solution of

$$\begin{cases} \partial_{tt} \Psi_{j,h} - \dfrac{1}{h^2} \left(\Psi_{j+1,h} - 2\Psi_{j,h} + \Psi_{j-1,h} \right) = 0, \\ \qquad\qquad\qquad\qquad (t,j) \in (0,T) \times \{1,\dots,N\}, \quad (4.76) \\ \Psi_{0,h}(t) = \Psi_{N+1,h}(t) = 0, \qquad t \in (0,T) \\ \Psi_h(T) = z_{T,h}, \ \partial_t \Psi_h(T) = 0. \end{cases}$$

4.4.2.3 Estimating the Difference $\int_0^1 y(T)\phi_T - \int_0^1 y_h(T)\phi_T$

First, since z_T solves the Laplace equation (4.66), $z_T \in H^2 \cap H_0^1(0,1)$ and

$$\|z_T\|_{H^2 \cap H_0^1} \simeq \|\phi_T\|_{L^2}.$$

Since $\phi_T \in L^2(0,1)$, using Theorems 4.1 and 4.3,

$$\|z_{T,h} - z_T\|_{H_0^1} \leq Ch\|\phi_T\|_{L^2}, \tag{4.77}$$

$$\left|\partial_x z_T(1) + \frac{z_{N,T,h}}{h}\right| \leq C\sqrt{h}\|\phi_T\|_{L^2}. \tag{4.78}$$

Hence using Proposition 3.8, we obtain

$$\sup_{t\in[0,T]} \|(\Psi_h, \partial_t\Psi_h, \partial_{tt}\Psi_h) - (\Psi, \partial_t\Psi, \partial_{tt}\Psi)\|_{H_0^1 \times L^2 \times H^{-1}}$$

$$+ \left\|\partial_x\Psi(t,1) + \frac{\Psi_{N,h}}{h}(t)\right\|_{L^2(0,T)} \leq Ch^{2/3}\|\phi_T\|_{L^2}. \tag{4.79}$$

We thus deduce that

$$\left|\int_0^T \partial_t v(t)\left(\frac{\Psi_{N,h}}{h} + \partial_x\Psi(t,1)\right)dt + \int_0^1 y^0(\partial_{tt}\Psi_h(0) - \partial_{tt}\Psi(0))\right.$$

$$\left. - \int_0^1 y^1(\partial_t\Psi_h(0) - \partial_t\Psi(0))\right| \leq Ch^{2/3}\|\phi_T\|_{L^2}\left(\|(y^0, y^1)\|_{H_0^1 \times L^2} + \|v\|_{H^1}\right).$$

According to Eqs. (4.69), (4.75), and the bound Eq. (4.78), this implies

$$\left|\int_0^1 (y_h(T) - y(T))\phi_T\right|$$

$$\leq C\left(\sqrt{h}|v(T)| + h^{2/3}(\|(y^0, y^1)\|_{H_0^1 \times L^2} + \|v\|_{H^1})\right)\|\phi_T\|_{L^2}.$$

Using now identity (4.59), we obtain the following result:

$$\|y_h(T) - y(T)\|_{L^2} \leq C\left(\sqrt{h}|v(T)| + h^{2/3}(\|(y^0, y^1)\|_{H_0^1 \times L^2} + \|v\|_{H^1})\right),$$

which implies that, if $v(T) = 0$,

$$\|y_h(T) - y(T)\|_{L^2} \leq Ch^{2/3}\left(\|(y^0, y^1)\|_{H_0^1 \times L^2} + \|v\|_{H^1}\right),$$

whereas otherwise

$$\|y_h(T) - y(T)\|_{L^2} \leq C\left(h^{2/3}\|(y^0, y^1)\|_{H_0^1 \times L^2} + \sqrt{h}\|v\|_{H^1}\right).$$

4.4.2.4 Conclusion

Note that all the above estimates hold uniformly for T in bounded intervals of time. This concludes the proof of Proposition 4.5. □

4.4.3 Convergence of $\partial_t y_h$

Proposition 4.6. *Under the assumptions of Theorem 4.7, taking (y_h^0, y_h^1) as in Eq. (4.58), we have the convergence (4.56).*

Proof. The proof of Proposition 4.6 closely follows the one of Proposition 4.5 and actually it is easier. We first begin by the following remark:

$$\|\partial_t y_h(T) - \partial_t y(T)\|_{H^{-1}} = \sup_{\substack{\phi_T \in H_0^1 \\ \|\phi_T\|_{H_0^1} \leq 1}} \left\{ \int_0^1 \partial_t y_h(T)\phi_T - \int_0^1 \partial_t y(T)\phi_T \right\}.$$

Hence we fix $\phi_T \in H_0^1(0, 1)$. We expand it in Fourier series:

$$\phi_T = \sum_{k=1}^{\infty} \hat{\phi}_k w^k, \quad \text{with } \|\phi_T\|_{H_0^1}^2 = \sum_{k=1}^{\infty} k^2 \pi^2 |\hat{\phi}_k|^2. \tag{4.80}$$

We thus introduce

$$\phi_{T,h} = \sum_{k=1}^{N} \hat{\phi}_k w^k.$$

Using the fact that $\partial_t y_h$ belongs to the span of the N-first Fourier modes,

$$\int_0^1 \partial_t y_h(T)\phi_T = \int_0^1 \partial_t y_h(T)\phi_{T,h}. \tag{4.81}$$

Hence we are reduced to show

$$\left| \int_0^1 \partial_t y(T)\phi_T - \int_0^1 \partial_t y_h(T)\phi_{T,h} \right|$$
$$\leq C h^{2/3} \left(\|(y^0, y^1)\|_{H_0^1 \times L^2} + \|v\|_{H^1} \right) \|\phi_T\|_{H_0^1}. \tag{4.82}$$

Again, we will express each of these quantities by an adjoint formulation and then relate the proof of Eq. (4.82) to convergence results for the adjoint system. Indeed,

$$\int_0^1 \partial_t y(T)\phi_T = \int_0^T v(t)\partial_x \varphi(t, 1)\, dt - \int_0^1 y^0 \partial_t \varphi(0) + \int_0^1 y^1 \varphi(0), \tag{4.83}$$

where φ solves

$$
\begin{cases}
\partial_{tt}\varphi - \partial_{xx}\varphi = 0, & (t,x) \in (0,T) \times (0,1), \\
\varphi(t,0) = \varphi(t,1) = 0, & t \in (0,T), \\
(\varphi(T), \partial_t \varphi(T)) = (\phi_T, 0).
\end{cases}
\tag{4.84}
$$

Then, introducing $\Phi(t) = \int_t^T \varphi(s)\,\mathrm{d}s$, we easily check that Φ solves

$$
\begin{cases}
\partial_{tt}\Phi - \partial_{xx}\Phi = 0, & (t,x) \in (0,T) \times (0,1), \\
\Phi(t,0) = \Phi(t,1) = 0, & t \in (0,T), \\
(\Phi(T), \partial_t \Phi(T)) = (0, -\phi_T).
\end{cases}
\tag{4.85}
$$

Besides, identity (4.83) then becomes

$$
\int_0^1 \partial_t y(T)\phi_T = \int_0^T \partial_t v(t)\partial_x \Phi(t,1)\,\mathrm{d}t + \int_0^1 y^0 \partial_{tt}\Phi(0) - \int_0^1 y^1 \partial_t \Phi(0). \tag{4.86}
$$

Similarly, we have

$$
\int_0^1 \partial_t y_h(T)\phi_{T,h} = -\int_0^T \partial_t v(t)\frac{\Phi_{N,h}}{h}(t)\,\mathrm{d}t + \int_0^1 y_h^0 \partial_{tt}\Phi_h(0) - \int_0^1 y_h^1 \partial_t \Phi_h(0), \tag{4.87}
$$

where Φ_h solves

$$
\begin{cases}
\partial_{tt}\Phi_{j,h} - \dfrac{1}{h^2}\left(\Phi_{j+1,h} + \Phi_{j-1,h} - 2\Phi_{j,h}\right) = 0, \\
\hspace{4cm} (t,j) \in (0,T) \times \{1,\ldots,N\}, \\
\Phi_{0,h}(t) = \Phi_{N+1,h}(t) = 0, \hspace{1cm} t \in (0,T), \\
(\Phi_h(T), \partial_t \Phi_h(T)) = (0, -\phi_{T,h}).
\end{cases}
\tag{4.88}
$$

Also remark that, since $\phi_{T,h}$ is formed by Fourier modes smaller than N, Φ_h has this same structure. Due to the orthogonality properties of the Fourier basis and the choice of the initial data in Eq. (4.58), we have

$$
\int_0^1 \partial_t y_h(T)\phi_{T,h} = -\int_0^T \partial_t v(t)\frac{\Phi_{N,h}}{h}(t)\,\mathrm{d}t + \int_0^1 y^0 \partial_{tt}\Phi_h(0) - \int_0^1 y^1 \partial_t \Phi_h(0). \tag{4.89}
$$

We are thus in the setting of Proposition 3.8 since $\phi_T \in H_0^1$ and one easily checks

$$
\left\|\phi_T - \phi_{T,h}\right\|_{L^2} \le Ch\|\phi_T\|_{H_0^1}.
$$

We thus obtain

$$
\sup_{t\in[0,T]} \left\|(\partial_t \Phi_h, \partial_{tt}\Phi_h) - (\partial_t \Phi, \partial_{tt}\Phi)\right\|_{L^2 \times H^{-1}} + \left\|\partial_x \Phi(t,1) + \frac{\Phi_{N,h}}{h}(t)\right\|_{L^2(0,T)}
$$
$$
\le Ch^{2/3}\|\phi_T\|_{H_0^1}. \tag{4.90}
$$

Then, using the identities (4.86) and (4.89), we get

$$\left| \int_0^1 \partial_t y(T) \phi_T - \int_0^T \partial_t y_h(T) \phi_{T,h} \right|$$

$$\leq Ch^{2/3} \|\phi_T\|_{H_0^1} \left(\|(y^0, y^1)\|_{H_0^1 \times L^2} + \|v\|_{H^1} \right). \tag{4.91}$$

Combined with Eq. (4.81), this easily yields Eq. (4.82).   □

4.4.4 More Regular Data

In this section, our goal is to explain what happens for smoother initial data (y^0, y^1) and v, for instance, for $(y^0, y^1) \in H^2 \cap H_0^1(0,1) \times H_0^1(0,1)$ and $v \in H^2(0,T)$ with $v(0) = \partial_t v(0) = 0$. More precisely, we are going to prove the following:

Theorem 4.8. *Let* $\ell_0 \in \{1,2\}$ *and fix* $(y^0, y^1) \in H_{(0)}^{\ell_0+1}(0,1) \times H_{(0)}^{\ell_0}(0,1)$ *and* $v \in H^{\ell_0+1}(0,T)$ *satisfying* $v(0) = \partial_t v(0) = 0$ *if* $\ell_0 = 1$, *or* $v(0) = \partial_t v(0) = \partial_{tt} v(0) = 0$ *if* $\ell_0 = 2$. *Let* (y_h^0, y_h^1) *be as in Eq.* (4.58) *and* y_h *the corresponding solution of Eq.* (4.7) *with Dirichlet boundary conditions* $v_h = v$.

Then there exists a constant $C > 0$ *independent of* $h > 0$ *and* $t \in [0,T]$ *such that:*
- *For the displacement* y_h, *for all* $t \in [0,T]$,

$$\|y_h(t) - y(t)\|_{L^2} \leq Ch^{2(\ell_0+1)/3} \left(\|(y^0, y^1)\|_{H_{(0)}^{\ell_0+1} \times H_{(0)}^{\ell_0}} + \|v\|_{H^{\ell_0+1}(0,T)} \right)$$

$$+ Ch^{1/2} |v(t)|. \tag{4.92}$$

- *For the velocity* $\partial_t y_h$, *for all* $t \in [0,T]$,

$$\|\partial_t y_h(t) - \partial_t y(t)\|_{H^{-1}} \leq Ch^{2(\ell_0+1)/3} \left(\|(y^0, y^1)\|_{H_{(0)}^{\ell_0+1} \times H_{(0)}^{\ell_0}} + \|v\|_{H^{\ell_0+1}(0,T)} \right)$$

$$+ Ch^{3/2} |\partial_t v(t)|. \tag{4.93}$$

Proof. The proof follows the one of Theorem 4.7.

Let us then focus on the convergence of the displacement and follow the proof of Proposition 4.5. We introduce $\phi_T \in L^2(0,1)$, z_T as in Eq. (4.66), Ψ the solution of the homogeneous wave equation (4.68) with initial data $(z_T, 0)$ and, similarly, $\phi_{T,h}$ as in Eq. (4.70), $z_{T,h}$ as in Eq. (4.74), and Ψ_h the solution of the discrete homogeneous wave equation (4.76) with initial data $(z_{T,h}, 0)$. Since $z_T \in H_{(0)}^2(0,1)$ and $\|z_T\|_{H_{(0)}^2} \simeq \|\phi_T\|_{L^2}$, applying (4.15), we get

$$\|z_{T,h} - z_T\|_{L^2} \leq Ch^2 \|\phi_T\|_{L^2}. \tag{4.94}$$

Proposition 3.8 then applies and yields

$$\|(\partial_t \Psi_h, \partial_{tt} \Psi_h) - (\partial_t \Psi, \partial_{tt} \Psi)\|_{H^{-\ell_0} \times H^{-\ell_0-1}} \le Ch^{2(\ell_0+1)/3} \|\phi_T\|_{L^2}.$$

In particular,

$$\left| \int_0^1 y^0 (\partial_{tt} \Psi_h(0) - \partial_{tt} \Psi(0)) - \int_0^1 y^1 (\partial_t \Psi_h(0) - \partial_t \Psi(0)) \right|$$
$$\le Ch^{2(\ell_0+1)/3} \|\phi_T\|_{L^2} \|(y^0, y^1)\|_{H^{\ell_0+1}_{(0)} \times H^{\ell_0}_{(0)}}. \tag{4.95}$$

According to identities (4.69) and (4.75), we shall then derive a convergence estimate on

$$\int_0^T \partial_t v \left(\partial_x \Psi(t, 1) + \frac{\Psi_{N,h}(t)}{h} \right) dt.$$

In order to do that, we write $\partial_t v = \int_0^t \partial_{tt} v$ and introduce

$$\xi(t) = \int_t^T \Psi(s) \, ds, \quad \xi_h(t) = \int_t^T \Psi_h(s) \, ds,$$

so that

$$\int_0^T \partial_t v \left(\partial_x \Psi(t, 1) + \frac{\Psi_{N,h}(t)}{h} \right) dt = \int_0^T \partial_{tt} v \left(\partial_x \xi(t, 1) + \frac{\xi_{N,h}(t)}{h} \right) dt.$$

Of course, ξ and ξ_h can be interpreted as solutions of continuous and discrete wave equations: ξ solves

$$\begin{cases} \partial_{tt} \xi - \partial_{xx} \xi = 0, & (t, x) \in (0, T) \times (0, 1) \\ \xi(t, 0) = \xi(t, 1) = 0, & t \in (0, T), \\ \xi(T) = 0, \ \partial_t \xi(T) = -z_T, \end{cases} \tag{4.96}$$

whereas ξ_h solves

$$\begin{cases} \partial_{tt} \xi_{j,h} - \dfrac{1}{h^2} \left(\xi_{j+1,h} - 2\xi_{j,h} + \xi_{j-1,h} \right) = 0, \\ & (t, j) \in (0, T) \times \{1, \ldots, N\}, \\ \xi_{0,h}(t) = \xi_{N+1,h}(t) = 0, & t \in (0, T), \\ \xi_h(T) = 0, \ \partial_t \xi_h(T) = -z_{T,h}. \end{cases} \tag{4.97}$$

Then, due to Eq. (4.94), the convergence results in Proposition 3.7 yield

$$\left\| \partial_x \xi(t, 1) + \frac{\xi_{N,h}(t)}{h} \right\|_{L^2(0,T)} \le Ch^{4/3} \|\phi_T\|_{L^2}.$$

This implies in particular that

$$\left| \int_0^T \partial_t v \left(\partial_x \Psi(t,1) + \frac{\Psi_{N,h}(t)}{h} \right) dt \right| \le C h^{4/3} \|\phi_T\|_{L^2} \|\partial_{tt} v\|_{L^2(0,T)}. \tag{4.98}$$

Hence, if $\ell_0 = 1$, i.e., $(y^0, y^1) \in H^2_{(0)}(0,1) \times H^1_{(0)}(0,1)$ and $v \in H^2(0,T)$ with $v(0) = \partial_t v(0) = 0$, combining Eqs. (4.95) and (4.98) in identities (4.69) and (4.75), we get

$$\|y_h(T) - y(T)\|_{L^2(0,1)} \le C h^{4/3} \left(\|(y^0, y^1)\|_{H^2_{(0)} \times H^1_{(0)}} + \|v\|_{H^2(0,T)} \right) + C h^{1/2} |v(T)|. \tag{4.99}$$

The Case $\ell_0 = 2$. In this case, $v \in H^3(0,T)$, we introduce $\zeta = \int_t^T \xi$ and $\zeta_h = \int_t^T \xi_h$, so that

$$\int_0^T \partial_t v \left(\partial_x \Psi(t,1) + \frac{\Psi_{N,h}(t)}{h} \right) dt = \int_0^T \partial_{tt} v \left(\partial_x \zeta(t,1) + \frac{\zeta_{N,h}(t)}{h} \right) dt. \tag{4.100}$$

Obviously, the function ζ can be characterized as the solution of a wave equation, namely,

$$\begin{cases} \partial_{tt} \zeta - \partial_{xx} \zeta = z_T, & (t,x) \in (0,T) \times (0,1) \\ \zeta(t,0) = \zeta(t,1) = 0, & t \in (0,T), \\ \zeta(T) = 0, \ \partial_t \zeta(T) = 0. \end{cases} \tag{4.101}$$

We thus introduce w_T solution of

$$\partial_{xx} w_T = z_T, \quad \text{on } (0,1), \quad w_T(0) = w_T(1) = 0, \tag{4.102}$$

so that

$$\tilde{\zeta} = \zeta - w_T$$

solves

$$\begin{cases} \partial_{tt} \tilde{\zeta} - \partial_{xx} \tilde{\zeta} = 0, & (t,x) \in (0,T) \times (0,1) \\ \tilde{\zeta}(t,0) = \tilde{\zeta}(t,1) = 0, & t \in (0,T), \\ \tilde{\zeta}(T) = w_T, \ \partial_t \tilde{\zeta}(T) = 0. \end{cases} \tag{4.103}$$

Doing that

$$\int_0^T \partial_{tt} v \, \partial_x \zeta(t,1) \, dt = \int_0^T \partial_{tt} v \, \partial_x \tilde{\zeta}(t,1) \, dt - \partial_x w_T(1) \partial_{tt} v(T). \tag{4.104}$$

Similar computations can be done for ζ_h. We thus obtain that

$$\int_0^T \partial_{tt} v \frac{\zeta_{N,h}(t)}{h} \, dt = \int_0^T \partial_{tt} v \frac{\tilde{\zeta}_{N,h}(t)}{h} \, dt - \frac{w_{N,T,h}}{h} \partial_{tt} v(T), \tag{4.105}$$

where $w_{T,h} = (\Delta_h)^{-1} z_{T,h}$ and $\tilde{\zeta}_h$ solves

$$
\begin{cases}
\partial_{tt} \tilde{\zeta}_{j,h} - \dfrac{1}{h^2} \left(\tilde{\zeta}_{j+1,h} - 2\tilde{\zeta}_{j,h} + \tilde{\zeta}_{j-1,h} \right) = 0, & \\
& (t,j) \in (0,T) \times \{1,\ldots,N\}, \\
\tilde{\zeta}_{0,h}(t) = \tilde{\zeta}_{N+1,h}(t) = 0, & t \in (0,T) \\
\tilde{\zeta}_h(T) = w_{T,h}, \ \partial_t \tilde{\zeta}_h(T) = 0.
\end{cases}
\tag{4.106}
$$

We now derive convergence estimates. Recall first that $z_T \in H^2_{(0)}(0,1)$ and the convergences (4.94). Since $z_T \in H^2_{(0)}$, setting $\tilde{z}_{T,h}$ its projection on the N-first Fourier modes, we have

$$
\left\| \tilde{z}_{T,h} - z_T \right\|_{L^2} \le Ch^2 \|z_T\|_{H^2_{(0)}} \le Ch^2 \|\phi_T\|_{L^2}.
\tag{4.107}
$$

Setting $\tilde{w}_{T,h} = (\Delta_h)^{-1} \tilde{z}_{T,h}$, Theorems 4.4 and 4.5 yield

$$
\begin{aligned}
\left\| w_T - \tilde{w}_{T,h} \right\|_{H^1_0} &\le & Ch^2 \|z_T\|_{H^2_{(0)}} \le Ch^2 \|\phi_T\|_{L^2}, \\
\left| \partial_x w_T(1) + \dfrac{\tilde{w}_{N,T,h}}{h} \right| &\le & Ch^2 \|z_T\|_{H^2_{(0)}} \le Ch^2 \|\phi_T\|_{L^2}.
\end{aligned}
\tag{4.108}
$$

According to the estimate (4.94), we thus have

$$
\left\| \tilde{z}_{T,h} - z_{T,h} \right\|_{L^2} \le Ch^2 \|z_T\|_{H^2_{(0)}} \le Ch^2 \|\phi_T\|_{L^2}.
$$

Using then estimate (4.21),

$$
\left| \dfrac{\tilde{w}_{N,T,h}}{h} - \dfrac{w_{N,T,h}}{h} \right| \le Ch^2 \|\phi_T\|_{L^2},
$$

and thus

$$
\left| \partial_x w_T(1) + \dfrac{w_{N,T,h}}{h} \right| \le Ch^2 \|\phi_T\|_{L^2}.
\tag{4.109}
$$

Besides, due to Eqs. (4.94) and (4.107),

$$
\left\| z_{T,h} - \tilde{z}_{T,h} \right\|_{L^2} \le Ch^2 \|\phi_T\|_{L^2},
$$

which readily implies

$$
\left\| w_{T,h} - \tilde{w}_{T,h} \right\|_{H^1_0} \le Ch^2 \|\phi_T\|_{L^2},
$$

and thus, by Eq. (4.108),

$$
\left\| w_{T,h} - w_T \right\|_{H^1_0} \le Ch^2 \|\phi_T\|_{L^2}.
$$

Using then Proposition 3.6,

$$
\left\| \partial_x \zeta(\cdot, 1) + \dfrac{\zeta_{N,h}}{h}(\cdot) \right\|_{L^2(0,T)} \le Ch^2 \|\phi_T\|_{L^2}.
\tag{4.110}
$$

Combined with the convergences (4.109) and (4.110), identities (4.100), (4.104), and (4.105) then imply

$$\left| \int_0^T \partial_t v \left(\partial_x \Psi(t,1) + \frac{\Psi_{N,h}(t)}{h} \right) dt \right|$$
$$\leq Ch^2 \|\phi_T\|_{L^2} \|\partial_{ttt} v\|_{L^2} + Ch^2 \|\phi_T\|_{L^2} |\partial_{tt} v(T)| \leq Ch^2 \|\phi_T\|_{L^2} \|v\|_{H^3}. \quad (4.111)$$

Combining Eqs. (4.95) and (4.111) in identities (4.69) and (4.75), we get Eq. (4.92) when $\ell_0 = 2$.

The proof of the estimate (4.93) on the rate of convergence for $\partial_t y_h$ relies on very similar estimates which are left to the reader.   □

4.5 Further Convergence Results

As a corollary to Theorems 4.6 and 4.7, we can give convergence results for *any* sequence of discrete initial data (y_h^0, y_h^1) and boundary data v_h satisfying

$$\lim_{h \to 0} \left\| (y_h^0, y_h^1) - (y^0, y^1) \right\|_{L^2 \times H^{-1}} = 0 \quad \text{and} \quad \lim_{h \to 0} \|v_h - v\|_{L^2(0,T)} = 0. \quad (4.112)$$

Proposition 4.7. *Let $(y^0, y^1) \in L^2(0,1) \times H^{-1}(0,1)$ and $v \in L^2(0,T)$. Then consider sequences of discrete initial data (y_h^0, y_h^1) and v_h satisfying Eq. (4.112). Then the solutions y_h of Eq. (4.7) with initial data (y_h^0, y_h^1) and boundary data v_h converge strongly in $C([0,T]; L^2(0,1)) \cap C^1([0,T]; H^{-1}(0,1))$ towards the solution y of Eq. (4.1) with initial data (y^0, y^1) and boundary data v as $h \to 0$.*

Proof. Similarly as in the proof of Proposition 3.5, this result is obtained by using the density of $H_0^1(0,T)$ in $L^2(0,T)$ and of $H_0^1(0,1) \times L^2(0,1)$ in $L^2(0,1) \times H^{-1}(0,1)$. We then use Theorem 4.7 for smooth solutions and the uniform stability results in Theorem 4.6 to obtain Proposition 4.7. Details of the proof are left to the reader.   □

Another important corollary of Theorem 4.7 is the fact that, if the initial data (y^0, y^1) belong to $H_0^1(0,1) \times L^2(0,1)$ and the Dirichlet data v lies in $H_0^1(0,T)$, *any* sequence of discrete initial (y_h^0, y_h^1) and Dirichlet data v_h satisfying

$$\left\| (y_h^0, y_h^1) - (y^0, y^1) \right\|_{L^2 \times H^{-1}} + \|v - v_h\|_{L^2(0,T)} \leq C_0 h^\theta, \quad (4.113)$$

for some constant C_0 uniform in $h > 0$ and $\theta > 0$, yield solutions y_h of Eq. (4.7) such that $y_h(T)$ approximates at a rate $h^{\min\{2/3, \theta\}}$ the state $y(T)$, where y is the continuous trajectory corresponding to initial data (y^0, y^1) and source term v.

Proposition 4.8. *Let $(y^0, y^1) \in H_0^1(0,1) \times L^2(0,1)$ and $v \in H_0^1(0,T)$ and consider sequences (y_h^0, y_h^1) and v_h satisfying Eq. (4.113).*

Denote by y_h (respectively y) the solution of Eq. (4.7) (resp. (4.1)) with initial data (y_h^0, y_h^1) (resp. (y^0, y^1)) and Dirichlet boundary data v_h, (resp. v).

Then the following estimates hold:

$$\|(y_h(T), \partial_t y_h(T)) - (y(T), \partial_t y(T))\|_{L^2 \times H^{-1}}$$
$$\leq Ch^{2/3} \left(\|(y^0, y^1)\|_{H_0^1 \times L^2} + \|v\|_{H_0^1(0,T)} \right) + CC_0 h^\theta. \tag{4.114}$$

Remark 4.4. In the convergence result Eq. (4.114), we keep explicitly the dependence in the constant C_0 coming into play in Eq. (4.113). In many situations, this constant can be chosen proportional to $\|(y^0, y^1)\|_{H_0^1 \times L^2} + \|v\|_{H_0^1(0,T)}$. In particular, in the control theoretical setting of Chap. 1 and its application to the wave equation in Sect. 1.7, this dependence on C_0 is important to derive Assumption 1 and more specifically estimate (1.29).

Proof. The proof follows the one of Proposition 3.7. The idea is to compare y with $\tilde{y}_h$, the solution of Eq. (4.7) constructed in Theorem 4.7 and then to compare $\tilde{y}_h$ and y_h by using Propositions 4.3 and 4.6.   □

Remark 4.5. Note that under the assumptions of Proposition 4.8, the trajectories y_h converge to y in the space $C([0,T]; L^2(0,1)) \cap C^1([0,T]; H^{-1}(0,1))$ with the rates (4.54)–(4.56) in addition to the error $C_0 h^\theta$.

Of course, Proposition 4.8 is based on the convergence result obtained in Theorem 4.7. Similar results can be stated based on Theorem 4.8, for instance:

Proposition 4.9. *Let $\ell_0 \in \{0, 1, 2\}$. Let $(y^0, y^1) \in H_{(0)}^{\ell_0+1}(0,1) \times H_{(0)}^{\ell_0}(0,1)$ and $v \in H_0^{\ell_0+1}(0,T)$ and consider sequences (y_h^0, y_h^1) and v_h satisfying Eq. (4.113).*

Let (y_h^0, y_h^1) as in Eq. (4.58) and y_h the corresponding solution of Eq. (4.7) with Dirichlet boundary conditions v_h.

Denote by y_h (respectively y) the solution of Eq. (4.7) (resp. Eq. (4.1)) with initial data (y_h^0, y_h^1) (resp. (y^0, y^1)) and Dirichlet boundary data v_h (resp. v).

Then the following estimates hold:

$$\|(y_h(T), \partial_t y_h(T)) - (y(T), \partial_t y(T))\|_{L^2 \times H^{-1}}$$
$$\leq Ch^{2(\ell_0+1)/3} \left(\|(y^0, y^1)\|_{H_{(0)}^{\ell_0+1} \times H_{(0)}^{\ell_0}} + \|v\|_{H_0^{\ell_0+1}(0,T)} \right) + CC_0 h^\theta. \tag{4.115}$$

Remark 4.6. Proposition 4.9 can then be slightly generalized for $\ell_0 \in [0,2]$ by interpolation.

4.6 Numerical Results

In this section, we present numerical simulations and evidences of Proposition 4.9. Since our main interest is in the non-homogeneous boundary condition, we focus on the case $(y^0, y^1) = (0,0)$ and $(y_h^0, y_h^1) = (0,0)$.

We fix $T = 2$. This choice is done for convenience to explicitly compute the solution y of Eq. (4.1) with initial data $(0,0)$ and source term v. Indeed, for $T = 2$, multiplying the equation (4.1) by φ solution of Eq. (3.2) with initial data $(\varphi^0, \varphi^1) \in H^1_0(0,1) \times L^2(0,1)$ and using the two-periodicity of the solutions of the wave equation (3.2), we obtain

$$\int_0^1 y(2,x)\varphi^1(x)\,dx - \int_0^1 \partial_t y(2,x)\varphi^0(x)\,dx = \int_0^2 v(t)\partial_x\varphi(t,1)\,dt.$$

Based on this formula, taking successively $(\varphi^0, \varphi^1) = (w^k, 0)$ and $(0, w^k)$ and solving explicitly the equation (3.2) satisfied by φ, we obtain

$$y(2) = \sum_k \left(\sqrt{2}(-1)^k \int_0^2 v(t)\sin(k\pi t)\,dt \right) w^k,$$

$$\partial_t y(2) = \sum_k \left(\sqrt{2}(-1)^{k+1} k\pi \int_0^2 v(t)\cos(k\pi t)\,dt \right) w^k.$$

We will numerically compute the reference solutions using these formulae by restricting the sums over $k \in \{1,\dots,N_{\text{ref}}\}$ for a large enough N_{ref}. We will choose $N_{\text{ref}} = 300$ for N varying between 50 and 200.

We then compute numerically the solution y_h of Eq. (4.7) with initial data $(y^0_h, y^1_h) = (0,0)$ and source term $v(t)$.

Of course, we also discretize the equation (4.7) in time. We do it in an explicit manner similarly as in Eq. (3.45). If y^k_h denotes the approximation of y_h solution of Eq. (4.7) at time $k\Delta t$, we solve

$$y^{k+1}_h = 2y^k_h - y^{k-1}_h - (\Delta t)^2 \Delta_h y^k_h - \left(\frac{\Delta t}{h}\right)^2 F^k, \quad F^k = \begin{pmatrix} 0 \\ \vdots \\ 0 \\ v(k\Delta t) \end{pmatrix}.$$

The time discretization parameter Δt is chosen such that the CFL condition is $\Delta t/h = 0.3$. With such low CFL condition, the effects of the time-discretization can be neglected.

We run the tests for several choices of v and for $N \in \{50,\dots,200\}$:

$$v_1(t) = \sin(\pi t)^3, \quad t \in (0,2), \qquad v_2(t) = \sin(\pi t)^2, \quad t \in (0,2),$$
$$v_3(t) = \sin(\pi t), \quad t \in (0,2), \qquad v_4(t) = t, \qquad\qquad t \in (0,2),$$
$$v_5(t) = t\sin(\pi t), \quad t \in (0,2).$$

In each case, we plot the L^2-norm of the error on the displacement and the H^{-1}-norm of the error on the velocity versus N in logarithmic scales: Fig. 4.2 corresponds to the data v_1. We then compute the slopes of the linear regression for the L^2-error on the displacement and for the H^{-1}-error on the velocity. We put all these data in Table 4.1.

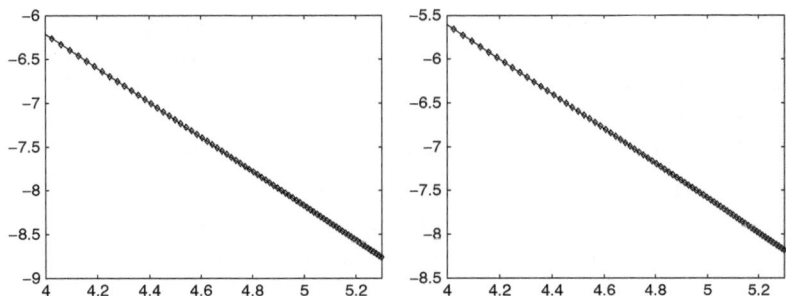

Fig. 4.2 Plots of the errors versus N in logarithmic scales for v_1. *Left*, the $L^2(0,1)$-error $\|y_h(T) - y(T)\|_{L^2}$ for $T = 2$: the slope of the linear regression is -1.96. *Right*, the $H^{-1}(0,1)$-error $\|\partial_t y_h(T) - \partial_t y(T)\|_{H^{-1}}$ for $T = 2$: the slope of the linear regression is -1.98.

Table 4.1 Numerical investigation of the convergence rates.

Data	Computed L^2 slope	Computed H^{-1} slope	Exp. L^2 slope	Exp. H^{-1} slope
v_1	-1.96	-1.98	-2	-2
v_2	-1.87	-1.70	$-5/3^-$	$-5/3^-$
v_3	-0.99	-0.95	-1^-	-1^-
v_4	-0.97	-0.95	$-1/2$	-1^-
v_5	-1.82	-1.47	$-5/3^-$	$-3/2$

Columns 2 and 3 give the slopes observed numerically (respectively, for the L^2-error on the displacement, for the H^{-1}-error on the velocity), whereas columns 4 and 5 provide the slopes (respectively, for the L^2-error on the displacement, for the H^{-1}-error on the velocity) expected from our theoretical results

Table 4.1 is composed of five columns. The first one is the data under consideration. The second and third ones, respectively, are the computed slopes of the linear regression of, respectively, the L^2-error on the displacement and for the H^{-1}-error on the velocity. The fourth and fifth columns are the rates expected from the analysis of the data v and Proposition 4.9:

- $v_1 \in H_0^3(0,2)$: we thus expect from Eq. (4.115) a convergence of the order of h^2. This is indeed what is observed numerically.
- v_2 is smooth but its boundary condition vanishes only up to order 1. Hence $v_2 \in H_0^{5/2-\varepsilon}(0,2)$ for all $\varepsilon > 0$ due to the boundary conditions. Using Remark 4.6, the expected slopes are $-5/3^-$, which is not far from the slopes computed numerically.
- The same discussion applies for v_3, which belongs to $H_0^{3/2-\varepsilon}(0,2)$ for all $\varepsilon > 0$. Hence the expected slopes are -1^-, which again are confirmed by the numerical experiments.
- v_4 almost belongs to $H_0^{3/2-\varepsilon}(0,2)$ except for what concerns its nonzero value at $t = 2$. But the value of v is an impediment for the order of convergence only for the displacement; see Theorem 4.8. We therefore expect a convergence of the L^2-norm of the error on the displacement like $\sqrt{h}$, whereas the convergence of the H^{-1}-norm of the error on the velocity is expected to go much faster, as h^{1^-}.

The numerical test indicates a good accuracy on the convergence of the H^{-1}-norm on the velocity error. The convergence of the L^2-norm of the displacement is better than expected.

- v_5 is smooth and satisfies $v_5(0) = \partial_t v_5(0) = 0$ and $v_5(2) = 0$ but $\partial_t v_5(2) \neq 0$. According to Theorem 4.8, we thus expect that the L^2-norm of the error on the displacement behaves as when v_5 belongs to $H_0^{5/2^-}(0,1)$, i.e., as $h^{5/3^-}$. However, the H^{-1}-norm of the error on the velocity should behave like $h^{3/2}$ according to Eq. (4.93). This is completely consistent with the slopes observed numerically.

In each case, the numerical results indicate good accuracy of the theoretical results derived in Theorem 4.8 and Proposition 4.9.

Chapter 5
Further Comments and Open Problems

5.1 Discrete Versus Continuous Approaches

We have developed the time-continuous and space discrete approaches for solving a control problem (and a data assimilation one) and we have proved that:

- The continuous approach works well for a limited number of iterations. In other words, the error between the continuous control and the approximated one decreases for a number of iterations. But, if one goes too far in the iteration process, beyond a threshold that theory predicts, the result can be completely misleading. Indeed, one eventually converges to a discrete control that is far away from the continuous one because of the high-frequency spurious oscillations. Thus, getting precise estimates on the threshold in the number of iterations is very important. But this is hard to do in practical applications since this requires, in particular, explicit bounds on the observability constants, something that is unknown in general and in particular for problems with variable coefficients, multidimensional problems with complex geometries, etc.

  The main advantage of the continuous approach is that it can be applied by simply combining the control theoretical results of the continuous model and the numerical convergence results for the initial–boundary value problem without any further study of the control theoretical properties of the numerical approximation scheme.

- The discrete approach yields good results after a given number of iterations (very close to the one of the continuous approach) and has the great advantage that the error keeps diminishing as the number of iterations increases. Thus there is no risk in going beyond any threshold in the number of iterations. However, guaranteeing that the discrete approach converges, contrarily to the continuous approach, requires the study of the control theoretical properties of the discrete systems and, in particular, the proof of a uniform observability result, uniformly with respect to the mesh size. This requires a good understanding of the dynamics of the solutions of numerical schemes and often careful filtering devices to eliminate the high-frequency spurious oscillations.

S. Ervedoza and E. Zuazua, *Numerical Approximation of Exact Controls for Waves*,
SpringerBriefs in Mathematics, DOI 10.1007/978-1-4614-5808-1_5,
© Sylvain Ervedoza and Enrique Zuazua 2013

- The main advantage of the discrete approach is that one may consider faster minimization algorithms, like conjugate gradient methods or more sophisticated ones, which converge often much faster. This justifies why the thorough study of uniform observability properties still is a major issue when numerically computing controls.

5.2 Comparison with Russell's Approach

The steepest descent algorithm applied in the continuous setting using the HUM approach leads to the following sequence of solutions of the adjoint problem

$$\varphi^k = \left(\sum_{j=0}^{k-1} (I - \rho \Lambda_T)^j\right)\rho y_0,$$

which, as k tends to infinity, approximates the solution of the adjoint system determining the exact control. Indeed, when letting $k \to \infty$, we get

$$\lim_{k \to \infty} \varphi^k = (I - (I - \rho \Lambda_T))^{-1}\rho y_0 = \Lambda_T^{-1} y_0.$$

Of course, this holds when the operator $(I - \rho \Lambda_T)$ is of norm strictly smaller than 1. This is precisely implied by the assumption that $\rho > 0$ is small enough and the fact that Λ_T is positive definite; see (1.47).

On the other hand, the approach developed in [9], inspired in Russell's iteration, which allows to get the control as a consequence of the stabilization property, leads to (see also [25] in the context of data assimilation)

$$\Psi_0 = \sum_{k \geq 0} (L_T)^k y_0,$$

where L_T is an operator of $\mathcal{L}(X)$ of norm strictly smaller than 1 and L_T is computed through the resolution of two wave equations (one forward and one backward) on $(0,T)$ (where $T \geq T^*$) with a damping term.

The numerical method proposed in [9] then follows the same strategy as our so-called continuous approach:

- Study the convergence of the sequence $\Psi_0^k = \sum_{j=0}^k (L_T)^j y_0$, in the spaces X and $\mathcal{D}(A)$. At this stage, the authors use that BB^* maps $\mathcal{D}(A^{3/2})$ into $\mathcal{D}(A)$.
- Approximate L_T by some discrete operator L_{Th} based on the natural approximations of the wave equation.
- Compare $\Psi_{0h}^k = \sum_{j=0}^k (L_{Th})^j y_{0h}$ with Ψ_0^k.
- Optimize the choice of k.

The method in [9] enters in the class of continuous methods. Note however that the continuous approach we proposed, inspired in HUM rather than on Russell's principle, does not require BB^* to map $\mathcal{D}(A^{3/2})$ into $\mathcal{D}(A)$.

In the continuous setting, the algorithm based on the time-reversal approach derived in [29] when recovering a source term is very close to Russell's approach: indeed, it corresponds to computing iterates of an operator of norm strictly smaller than one and deduced from the resolution of two dissipative wave equations. In that context, one easily understands that the approach in [25] enters the framework of the continuous approach based on [29].

5.3 Uniform Discrete Observability Estimates

The discrete approach relies in an essential manner upon the uniform observability estimates (1.37) of the semi-discrete approximations of the continuous model, i.e., Assumption 3, which, as we have said, is not an easy task to prove in practice.

In particular, to our knowledge, there are only few results which hold in general geometric settings and for regular finite-element method (not necessarily on uniform meshes), namely the ones in [12, 41]. However, these two works do not yield estimates on the time under which uniform observability holds. This is due to their strategy, based on resolvent estimates as a characterization of observability; see for instance [40]. The scale of filtering employed in these works to guarantee uniform discrete observability estimates is very likely not optimal. Its improvement is an interesting open problem.

Therefore, getting uniform observability estimates still is a challenging issue when considering general geometric setting guaranteeing the observability inequality (1.5) of the continuous model, in particular with respect to the time and the scale of filtering required for guaranteeing uniform discrete observability estimates.

5.4 Optimal Control Theory

Optimal control problems and the design of feedback control systems are topics that are closely related to the questions we have analyzed. Similarly to the numerical algorithms for exact control problem we studied here, we could also address the problem of numerically computing feedback control operators. As one could expect, getting discrete optimal feedback controls which converge to the continuous one usually requires the so-called uniform stabilizability property (see [19, 30, 32]), ensuring that the exponential decay rate of the energy of the solutions, both continuous and discrete, is bounded from below uniformly with respect to the mesh-size parameter. This issue is very closely related to the uniform discrete observability estimates (1.37). In [14], following the approach of [26], we explained how discrete observability inequalities can be transferred into uniform stabilizability results for the corresponding damped equations by the addition of a suitable numerical viscosity. This should provide convergent approximations of optimal feedback operators, as it has been done in [44].

However, to our knowledge, getting explicit rates on the convergence of these feedback controllers is an open problem.

5.5 Fully Discrete Approximations

Our approach is very general and can also be applied to fully discrete systems under very similar assumptions. For instance, one can formulate the analogs of Assumptions 1 and 2 that take into account the required convergence properties of the fully discrete numerical approximation scheme, whereas Assumption 3 consists of a uniform (with respect to the space-time mesh-size parameters) observability result for the fully discrete systems.

Note that, according to the results in [18], the corresponding fully discrete version of Assumption 3, which reads as uniform observability estimates for the fully discrete system, can be deduced as a consequence of the time-continuous (and space discrete) analogs.

References

1. D. Auroux, J. Blum, Back and forth nudging algorithm for data assimilation problems. C. R. Math. Acad. Sci. Paris **340**(12), 873–878 (2005)
2. G.A. Baker, J.H. Bramble, Semidiscrete and single step fully discrete approximations for second order hyperbolic equations. RAIRO Anal. Numér. **13**(2), 75–100 (1979)
3. C. Bardos, G. Lebeau, J. Rauch, Sharp sufficient conditions for the observation, control and stabilization of waves from the boundary. SIAM J. Contr. Optim. **30**(5), 1024–1065 (1992)
4. S.C. Brenner, L.R. Scott, The mathematical theory of finite element methods, in *Texts in Applied Mathematics*, vol. 15 (Springer, New York, 1994)
5. N. Burq, P. Gérard, Condition nécessaire et suffisante pour la contrôlabilité exacte des ondes. C. R. Acad. Sci. Paris Sér. I Math. **325**(7), 749–752 (1997)
6. C. Castro, S. Micu, Boundary controllability of a linear semi-discrete 1-d wave equation derived from a mixed finite element method. Numer. Math. **102**(3), 413–462 (2006)
7. C. Castro, S. Micu, A. Münch, Numerical approximation of the boundary control for the wave equation with mixed finite elements in a square. IMA J. Numer. Anal. **28**(1), 186–214 (2008)
8. P.G. Ciarlet, in *Introduction à l'analyse numérique matricielle et àl' optimisation*. Collection Mathématiques Appliquées pour la Maîtrise [Collection of Applied Mathematics for the Master's Degree] (Masson, Paris, 1982)
9. N. Cîndea, S. Micu, M. Tucsnak, An approximation method for exact controls of vibrating systems. SIAM J. Contr. Optim. **49**(3), 1283–1305 (2011)
10. J.-M. Coron, S. Ervedoza, O. Glass, Uniform observability estimates for the 1-d discretized wave equation and the random choice method. Comptes Rendus Mathematique **347**(9–10), 505–510 (2009)
11. B. Dehman, G. Lebeau, Analysis of the HUM control operator and exact controllability for semilinear waves in uniform time. SIAM J. Contr. Optim. **48**(2), 521–550 (2009)

S. Ervedoza and E. Zuazua, *Numerical Approximation of Exact Controls for Waves*,   119
SpringerBriefs in Mathematics, DOI 10.1007/978-1-4614-5808-1,

12. S. Ervedoza, Spectral conditions for admissibility and observability of wave systems: applications to finite element schemes. Numer. Math. **113**(3), 377–415 (2009)
13. S. Ervedoza, Observability properties of a semi-discrete 1D wave equation derived from a mixed finite element method on nonuniform meshes. ESAIM Contr. Optim. Calc. Var. **16**(2), 298–326 (2010)
14. S. Ervedoza, E. Zuazua, Uniformly exponentially stable approximations for a class of damped systems. J. Math. Pures Appl. **91**, 20–48 (2009)
15. S. Ervedoza, E. Zuazua, A systematic method for building smooth controls for smooth data. Discrete Contin. Dyn. Syst. Ser. B **14**(4), 1375–1401 (2010)
16. S. Ervedoza, E. Zuazua, The wave equation: control and numerics, in *Control of Partial Differential Equations*, ed. by P.M. Cannarsa, J.M. Coron. Lecture Notes in Mathematics, CIME Subseries (Springer, New York, 2012), pp. 245–340
17. S. Ervedoza, E. Zuazua, *Propagation, Observation and Numerical Approximations of Waves*. Book in preparation.
18. S. Ervedoza, C. Zheng, E. Zuazua, On the observability of time-discrete conservative linear systems. J. Funct. Anal. **254**(12), 3037–3078 (2008)
19. J.S. Gibson, A. Adamian, Approximation theory for linear-quadratic-Gaussian optimal control of flexible structures. SIAM J. Contr. Optim. **29**(1), 1–37 (1991)
20. R. Glowinski, Ensuring well-posedness by analogy: Stokes problem and boundary control for the wave equation. J. Comput. Phys. **103**(2), 189–221 (1992)
21. R. Glowinski, C.H. Li, On the numerical implementation of the Hilbert uniqueness method for the exact boundary controllability of the wave equation. C. R. Acad. Sci. Paris Sér. I Math. **311**(2), 135–142 (1990)
22. R. Glowinski, W. Kinton, M.F. Wheeler, A mixed finite element formulation for the boundary controllability of the wave equation. Int. J. Numer. Meth. Eng. **27**(3), 623–635 (1989)
23. R. Glowinski, C.H. Li, J.-L. Lions, A numerical approach to the exact boundary controllability of the wave equation. I. Dirichlet controls: description of the numerical methods. Japan J. Appl. Math. **7**(1), 1–76 (1990)
24. R. Glowinski, J.-L. Lions, J. He, in *Exact and Approximate Controllability for Distributed Parameter Systems: A Numerical Approach*. Encyclopedia of Mathematics and its Applications, vol. 117 (Cambridge University Press, Cambridge, 2008)
25. G. Haine, K. Ramdani, Reconstructing initial data using observers: error analysis of the semi-discrete and fully discrete approximations. Numer. Math. **120**(2), 307–343 (2012)
26. A. Haraux, Une remarque sur la stabilisation de certains systèmes du deuxième ordre en temps. Port. Math. **46**(3), 245–258 (1989)
27. L.F. Ho, Observabilité frontière de l'équation des ondes. C. R. Acad. Sci. Paris Sér. I Math. **302**(12), 443–446 (1986)
28. J.A. Infante, E. Zuazua, Boundary observability for the space semi discretizations of the 1-d wave equation. Math. Model. Num. Ann. **33**, 407–438 (1999)

29. K. Ito, K. Ramdani, M. Tucsnak, A time reversal based algorithm for solving initial data inverse problems. Discrete Contin. Dyn. Syst. Ser. S **4**(3), 641–652 (2011)

30. F. Kappel, D. Salamon, An approximation theorem for the algebraic Riccati equation. SIAM J. Contr. Optim. **28**(5), 1136–1147 (1990)

31. V. Komornik, in *Exact controllability and stabilization: The multiplier method.* RAM: Research in Applied Mathematics (Masson, Paris, 1994)

32. I. Lasiecka, Galerkin approximations of infinite-dimensional compensators for flexible structures with unbounded control action. Acta Appl. Math. **28**(2), 101–133 (1992)

33. I. Lasiecka, R. Triggiani, Regularity of hyperbolic equations under L_2 $(0, T; L_2(\Gamma))$-Dirichlet boundary terms. Appl. Math. Optim. **10**(3), 275–286 (1983)

34. I. Lasiecka, J.-L. Lions, R. Triggiani, Nonhomogeneous boundary value problems for second order hyperbolic operators. J. Math. Pures Appl. (9) **65**(2), 149–192 (1986)

35. J.-L. Lions, in *Contrôle des systèmes distribués singuliers.* Lectures at Collège de France, 1982 (Gauthier Villars, Paris, 1983)

36. J.-L. Lions, in *Contrôlabilité exacte, Stabilisation et Perturbations de Systèmes Distribués. Tome 1. Contrôlabilité exacte*, RMA, vol. 8 (Masson, Paris, 1988)

37. J.-L. Lions, Exact controllability, stabilization and perturbations for distributed systems. SIAM Rev. **30**(1), 1–68 (1988)

38. F. Macià, The effect of group velocity in the numerical analysis of control problems for the wave equation, in *Mathematical and Numerical Aspects of Wave Propagation—WAVES 2003* (Springer, Berlin, 2003), pp. 195–200

39. S. Micu, Uniform boundary controllability of a semi-discrete 1-D wave equation. Numer. Math. **91**(4), 723–768 (2002)

40. L. Miller, Controllability cost of conservative systems: resolvent condition and transmutation. J. Funct. Anal. **218**(2), 425–444 (2005)

41. L. Miller, Resolvent conditions for the control of unitary groups and their approximations. J. Spectra. Theor. **2**(1), 1–55 (2012).

42. M. Negreanu, E. Zuazua, Convergence of a multigrid method for the controllability of a 1-d wave equation. C. R. Math. Acad. Sci. Paris **338**(5), 413–418 (2004)

43. M. Negreanu, A.-M. Matache, C. Schwab, Wavelet filtering for exact controllability of the wave equation. SIAM J. Sci. Comput. **28**(5), 1851–1885 (electronic) (2006)

44. K. Ramdani, T. Takahashi, M. Tucsnak, Uniformly exponentially stable approximations for a class of second order evolution equations—application to LQR problems. ESAIM Contr. Optim. Calc. Var. **13**(3), 503–527 (2007)

45. J. Rauch, On convergence of the finite element method for the wave equation. SIAM J. Numer. Anal. **22**(2), 245–249 (1985)

46. P.-A. Raviart, J.-M. Thomas, in *Introduction à l'analyse numérique des équations aux dérivées partielles.* Collection Mathématiques Appliquées pour la Maitrise. [Collection of Applied Mathematics for the Master's Degree] (Masson, Paris, 1983)

47. D.L. Russell, Controllability and stabilizability theory for linear partial differential equations: recent progress and open questions. SIAM Rev. **20**(4), 639–739 (1978)
48. L.N. Trefethen, Group velocity in finite difference schemes. SIAM Rev. **24**(2), 113–136 (1982)
49. M. Tucsnak, G. Weiss, in *Observation and Control for Operator Semigroups*. Birkäuser advanced texts, vol. XI (Springer, Berlin, 2009)
50. R. Vichnevetsky, J.B. Bowles, in *Fourier analysis of numerical approximations of hyperbolic equations*. SIAM studies in applied mathematics, vol. 5 (SIAM, Philadelphia, 1982). With a foreword by G. Birkhoff
51. E. Zuazua, Boundary observability for the finite-difference space semi-discretizations of the 2-D wave equation in the square. J. Math. Pures Appl. (9) **78**(5), 523–563 (1999)
52. E. Zuazua, Propagation, observation, and control of waves approximated by finite difference methods. SIAM Rev. **47**(2), 197–243 (electronic) (2005)